Information and Organization Design Series

Volume 8

For further volumes:
http://www.springer.com/series/6126

Anne Bøllingtoft · Dorthe Døjbak Håkonsson ·
Jørn Flohr Nielsen · Charles C. Snow · John Ulhøi
Editors

New Approaches to Organization Design

Theory and Practice of Adaptive Enterprises

Springer

Editors
Anne Bøllingtoft
Aarhus School of Business
Aarhus University
Denmark
anns@asb.dk

Dorthe Døjbak Håkonsson
Aarhus School of Business
Aarhus University
Denmark
dod@asb.dk

Jørn Flohr Nielsen
Aarhus School of Business
Aarhus University
Denmark
jflohr@asb.dk

Charles C. Snow
The Pennsylvania State University
USA
csnow@psu.edu

John Ulhøi
Aarhus School of Business
Aarhus University
Denmark
jpu@asb.dk

ISBN 978-1-4419-0626-7 e-ISBN 978-1-4419-0627-4
DOI 10.1007/978-1-4419-0627-4
Springer Dordrecht Heidelberg London New York

Library of Congress Control Number: 2009931578

Printed on acid-free paper

Springer is part of Springer Science+Business Media (www.springer.com)

On August 12 2009, Richard Burton will be celebrating his 70th birthday. Such a milestone provides a natural occasion to look back and take stock of one's work. Rich, however, would be the last person to draw attention to himself, to loudly proclaim his own accomplishments. And yet he has had a profound impact on our field.

With his frequent visits to Denmark (starting in 1974), Rich was one of the natural founders of the International Workshop on Organization Design. We are proud to take the publication of this book, originating from this conference, as an excellent opportunity to celebrate Rich and his major accomplishments.

Rich's numerous books and articles have been published in the major journals within our field, and have had a far-reaching and substantial impact on the field of Organization Design. His research has advanced our understanding of organization theory and design, not the least by the use of simulation techniques. Rich has helped create and improve the journals Organization Science, Computational and Mathematical Organization Theory, and Management Science. He has helped found institutions such as the College on Organization Science, the EIASM conference on Organization Design, and the Organization Science Winter Conference. In many ways, Rich has been the catalyst in the formation of our international research community.

But above all, Rich has been and remains a friend and mentor to many of us, an exemplar of a gentleman and a scholar.

This book is dedicated to Rich.

Contents

Part I Toward New Organizational Forms

1 **Blade.Org: A Collaborative Community of Firms** 3
Charles C. Snow, Doreen R. Strauss, and Christopher Lettl

2 **Network-Level Task and the Design of Whole Networks: Is There a Relationship?** 23
Patrick Kenis, Keith G. Provan, and Peter M. Kruyen

Part II Dynamics of Adaptation and Change

3 **Organizational Trade-Offs and the Dynamics of Adaptation in Permeable Structures** 43
Stephan Billinger and Nils Stieglitz

4 **Unpacking Dynamic Capability: A Design Perspective** 61
Deborah E. M. Mulders and A. Georges L. Romme

5 **Predicting Organizational Reconfiguration** 79
Timothy N. Carroll and Samina Karim

6 **Embedding Virtuality into Organization Design Theory: Virtuality and Its Information Processing Consequences** 99
Kent Wickstrøm Jensen, Dorthe Døjbak Håkonsson, Richard M. Burton, and Børge Obel

Part III Fit and Performance

7 **Learning-Before-Doing and Learning-in-Action: Bridging the Gap Between Innovation Adoption, Implementation, and Performance** 123
Eitan Naveh, Ofer Meilich, and Alfred Marcus

8 Underfits Versus Overfits in the Contingency Theory of Organizational Design: Asymmetric Effects of Misfits on Performance 147
Peter Klaas and Lex Donaldson

Index 169

Contributors

Stephan Billinger Strategic Organization Design Group, Department of Marketing and Management, University of Southern Denmark, Odense, Denmark, sbi@sam.sdu.dk

Richard M. Burton Fuqua School of Business, Duke University, Box 90120, Durham, NC 27708-0120, USA, rmb2@duke.edu

Timothy N. Carroll Moore School of Business, University of South Carolina, Columbia, SC 29208, USA, carroll@moore.sc.edu

Lex Donaldson Australian School of Business, University of New South Wales, Sydney, NSW 2052, Australia, lexd@agsm.edu.au

George P. Huber McCombs School of Business, University of Texas at Austin, Austin, TX, USA, george.huber@mccombs.utexas.edu

Dorthe Døjbak Håkonsson Aarhus School of Business, University of Aarhus, Haslegårdsvej 10, 8210 Arhus V, Denmark, dod@asb.dk

Kent Wickstrøm Jensen Department of Entrepreneurship and Relationship Management, University of Southern Denmark, Engstien 1, 6000, Kolding, Denmark, kwj@sam.sdu.dk

Samina Karim School of Management, Boston University, Boston, MA 02215, USA, samina@bu.edu

Patrick Kenis TiasNimbas Business School, Tilburg University, Tilburg, The Netherlands, p.kenis@tiasnimbas.edu

Peter Klaas Vestas Assembly A/S, Rymarken 2, DK 8210 Aarhus V, Denmark, klaas@mail.dk

Peter M. Kruyen Department of Organisation Studies & Tilburg School of Politics and Public Administration, Tilburg University, Tilburg, The Netherlands, p.m.kruyen@uvt.nl

Christopher Lettl Department of Marketing and Statistics, Aarhus School of Business, University of Aarhus, DK-8210 Aarhus V, Denmark, lettl@asb.dk

Alfred Marcus Carlson School of Management, University of Minnesota, Minneapolis, MN, 55455, USA, amarcus@csom.umn.edu

Ofer Meilich College of Business Administration, California State University, San Marcos, CA 92096, USA, meilich@csusm.edu

Deborah E. M. Mulders Eindhoven University of Technology, Eindhoven, The Netherlands, d.e.m.mulders@tue.nl

Eitan Naveh Faculty of Industrial Engineering and Management, Technion – Israel Institute of Technology, Haifa 32000, Israel, naveh@ie.technion.ac.il

Børge Obel Aarhus School of Business, University of Aarhus, Fuglesangs Alle 4, 8210 Aarhus V, Denmark, bo@asb.dk

Keith G. Provan Eller College of Management, University of Arizona, Tucson, AZ, USA; Department of Organisation Studies, Tilburg University, Tilburg, The Netherlands, kprovan@eller.arizona.edu

A. Georges L. Romme Eindhoven University of Technology, Eindhoven, The Netherlands, a.g.l.romme@tm.tue.nl

Charles C. Snow Department of Management and Organization, Smeal College of Business, The Pennsylvania State University, University Park, PA 16802 USA, csnow@psu.edu

Nils Stieglitz Strategic Organization Design Group, Department of Marketing and Management, University of Southern Denmark, Odense, Denmark, nst@sam.sdu.dk

Doreen R. Strauss Schreyer Honors College, The Pennsylvania State University, University Park, PA 16802 USA, doreen.r.strauss@accenture.com

Introduction: Use of Theory in Organization Design Research

George P. Huber

This volume had its origins in a conference on New Organization and Design Approaches: Anticipating the Future, sponsored by the Aarhus School of Business (ASB), Aarhus University, the CORE research center at ASB, and the Danish Social Science Research Council. In this introduction I begin by examining prominent organization theories in terms of how they might be extended to be more valid in today's and tomorrow's organizational environments and how they might be extended to be more useful to organization design research and practice. Afterward, I discuss the research studies reported in this book's chapters and comment on the use or non-use of organization theories by the researchers.

The State of Organization Theories

Here, by an *organization theory,* I mean a cohesive set of beliefs held in the organization science community that are, or could be, expressed as related propositions specifying relationships among organizationally relevant variables (cf. Kerlinger, 1986: 9). The current validity and/or usefulness of some of the most well established organization theories is in question.

> A number of paradigms for the study of organizations were elaborated during the mid-1970s, including transaction cost economics, resource dependence theory, organizational ecology, new institutional theory, and agency theory in financial economics. These approaches reflected the dominant trends of the large corporation of their time; increasing concentration, diversification, and bureaucratization. However, subsequent shifts in organizational boundaries, the increased use of alliances and network forms, and the expanding role of financial markets in shaping organizational decision making all make normal science driven by the internally derived questions from these paradigms less fruitful (Davis and Marquis 2005: 332; see also Daft and Lewin 1990).

The suggestions of Dorthe Døjbak Håkonsson were very useful in the preparation of this work.

George P. Huber (✉)
McCombs School of Business, University of Texas at Austin, Austin, TX, USA
e-mail: george.huber@mccombs.utexas.edu

> As most theories of organizational design derive from research conducted before 1980, one could argue that organization design is currently stuck in a theoretical rut. New kinds of organizations have emerged over the last 30 years; communication technologies have revolutionized organizations; knowledge-based activities are now central to working lives, and educational levels have risen. Although these changes must directly influence organization design, little research has explored these influences (Dunbar and Starbuck 2004: 497; see also Daft and Lewin 1993).

Given the deficiencies noted by Davis and Marquis (2005) we might question whether the theories established in the 1970s are now being used by organization theory researchers. *Are organization theory researchers using the more established organization theories in their current research?*

It seems that, on the whole, they are not. Walsh, Meyer, and Schoonhoven (2006), for example, report that of the 429 submissions to the Organization and Management Theory Division for presentation at the Academy of Management's 2005 annual meeting, "the percentage of papers submitted in each of the established theoretical categories (as reported by the authors of the papers) were as follows: institutional theory, 25.4%; network theory, 16.8%; population ecology, 6.7%; agency theory, 4.5%; resource dependence theory, 3.9%; transaction cost theory, 3.4%; contingency theory, 2.5%; and stakeholder theory, 2.5%. 'None of the above' accounted for 56% of the papers submitted" (2006: 658). Whatever the reasons, and there may be good ones, it appears that more than half of the active organization theory researchers in this large convenience sample were not using an established organization theory to underpin their current work.

Dunbar and Starbuck (2004) change the focus just a bit, from organization theories to organization *design* theories (by which, I will assume, they mean organization theories from which organization design guidelines can be inferred). Relevant to this volume are two more questions: *Are current organization theories useful to organization design researchers? Are organization design researchers using organization theories in their current work on organization design?* It seems that there are good reasons why an organization design researcher might not use current organization theories. *One* is that some theories are useful for understanding the effects of organizational environments on organizations but do not readily lend themselves to the development of organization design guidelines (population ecology and institutional theory come to mind). A *second* reason is that several of the theories were developed in earlier eras, when organizational environments were quite different from those of the present or the future. Thus the validity of the theory may be low in the future-focused temporal domain of interest to the organization design researcher. A *third* reason is that, as is necessarily the case, not all organization design researchers can be aware of all of the changes that are developing in all of the theories and, consequently, are not drawing on the more temporally relevant versions of the theories.

Extending the Usefulness of Some Prominent Organization Theories

Below I discuss some prominent organization theories that are congruent with the positivistic and empirical verification characteristics of the organization theory field's paradigm (see Aldrich and Ruef 2006; Daft 2007; and Scott and Davis 2007, for descriptions of these theories). I do this partly to re-familiarize readers with those theories that they may not have examined recently, but more so to suggest current or needed developments if the theories are to be more relevant to the design of organizations for current and future organizational environments. With regard to these environments, I hold that *future organizational environments will be characterized by* (1) increases in the number and effectiveness of information, manufacturing, and transportation technologies, (2) more and increasing environmental complexity, (3) more and increasing environmental dynamism, and (4) more and increasing environmental competitiveness. (For rationale and data supporting the existence of these four characteristics in future organizational environments, see Huber 2004. For an examination of the probable consequences of these environmental characteristics on the size, founding rates, performance, and survival of future organizations, see Huber forthcoming).

In the following analyses, I frequently speak of an organization's nature or circumstances as dependent or outcome variables. By an organization's *nature*, I mean the sum total of the attributes of its leadership, strategy, core technology, structure, employees, culture, and practices, and also the propensities and properties associated with these features, such as the organization's inclinations and competences. By an organization's *circumstances,* I mean its size, maturity, performance, status with regard to survival, and the current direction and speed of change of these variables.

Population ecology theory asserts that the suitability of the attributes of the organizations in a population, relative to the attributes of organizations in competing populations, determines the nature and circumstances (generally the survival or demise) of the population. In their early formulation of the theory, population ecologists viewed organizational adaptation to changing environmental conditions as highly unlikely due to inertia forces. Many might still so believe, but some more recent population ecology work does examine conditions where adaptation might be sufficient to maintain survival (Barnett and Carroll 1995; Dobrey et al. 2002). More such studies would be very useful to those organization design researchers who focus on adaptive redesign of an organization in the face of environmental change. Population ecology studies investigating whether the relationship between a particular organizational attribute and organizational survival actually has not changed when business press coverage indicates that it has changed (or vice versa) would also be useful. Neither of these two types of studies is commonplace in the population ecology literature, but would be welcome developments for both population ecologists and organization design researchers, especially in the face of more frequent and rapid environmental changes.

Institutional theory holds that institutional forces are a major influence on the nature of an organization (or of a population). Some institutional theorists are recognizing that organizations can influence institutions and resist institutional forces, that the relationship is not just one way (Haunschild and Chandler 2008). Studies investigating the circumstances and methods where this is possible would be very useful to those organization design researchers who focus on adaptation strategies for coping with environmental change. Further, when organizational environments change rapidly, organizations might change more rapidly than can institutional forces (given that values and norms are slow to change), providing a window in which organizations are less institutionally constrained. The lack of regulation of the Internet might be an example of such a relatively unconstrained situation. Studies examining the relative speed of change would be welcome additions to the institutional theory literature and to policy makers, executives, and organization design researchers.

Network theory states that an organization's internal and external network structures and the attributes of the network nodes determine the organization's performance. Network theorists have been generating findings useful to organization designers at a relatively rapid rate and covering a broad scope of issues pertinent to future organizational environments (e.g., Provan and Sebastian 1998; Rosenkopf and Padula 2008; Soda et al. 2004; Uzzi 1996). Thus organization designers would appear to be well served by network theorists. What would seem to be useful to network theorists and organization design researchers alike at this time would be studies that examine the boundary conditions of existing studies and review pieces that integrate empirical findings into formal theories and design guidelines that can be used by organization design researchers.

Transaction cost theory focuses on the decisions and contracts regarding where the organization's work is done – inside or outside of the organization. It asserts that the relative costs and risks associated with contacting work out versus having the work done within the organization cause the boundaries of an organization to be what they are. As the Walsh, Meyer, and Schoonhoven (2006) study notes, little work is being done in transactions cost theory. Advances in science and expansions in the variety and number of technologies have contributed to the need for organizations to specialize, while advances in communication and transportation technologies have contributed to the ability of organizations to provide their specialized products and services to other organizations efficiently. The consequence of these changes to organizational environments is that many firms have reduced their scope and many new firms choose to have work done outside the organization if it is not tightly linked to their core competence. Such firms, when they grow, grow either by adopting the strategy of selling their specialized products or services to multiple customers or by adopting the strategy of participating as specialists in multiple networks. It seems that transaction cost theorists could contribute more to designing organizations for the future by drawing on their own theory (1) to study the conditions under which firms can most advantageously adopt which of these, or other, strategies and designs and (2) to identify leading indicators useful for identifying when firms should change their boundaries.

Resource dependence theory claims that the organization's actions and level of success in attaining power over those organizations that possess resources on which the organization depends are what determine the circumstances of an organization. The purpose of these actions is to reduce dependencies and uncertainties about the availability of resources. Advances in information, manufacturing, and transportation technologies provide all organizations with more options and thus make it more difficult for an organization to maintain control over an unwillingly dependent organization. Perhaps as a result, it appears that in recent years many firms have chosen to reduce dependencies and uncertainties through cooperative relationships (i.e., alliances) rather than through power dominance. If resource dependence theorists drew on their own theory to investigate more intensely the variables which favor the use of collaboration versus dominance, it seems they could make important contributions to organization design theory and to the design of organizations and networks of organizations operating in future organizational environments.

Strategic choice theory asserts that the judgments and preferences of the organization's dominant coalition, subject to the coalition's interpretation of strength of the constraints posed by the organization's internal and external environments and to its success in getting its preferred organizational attributes emplaced and maintained, are what determine the nature and circumstances of an organization. While strategic choice theory has influenced other theories (Child 1997), advances to the theory itself have been largely in the form of works concerning the upper echelons of organizations (see Hambrick's 2007 review of this literature), particularly the echelon's discretion. It appears that strategic choice theorists could contribute even more to the organization design literature if they provided more studies of the factors contributing to the dominant coalition's effectiveness in making organization design choices, especially in the more complex, dynamic, and competitive organizational environments that organizations will encounter in the future.

The *resource-based view of the firm* (RBV) holds that the firm's ability to create and implement strategies that develop and exploit those of its resources which are valuable, rare, not readily imitated, and not substitutable explains the relative competitiveness of a firm. A currently active area of research and an important extension of the RBV for future environments is the concept of *dynamic capabilities*, defined as "the firm's ability to integrate, build, and reconfigure internal and external competences to address rapidly changing conditions" (Teece, et al., 1997: 516). Examples given by Eisenhardt and Martin (2000) are "product development, strategic decision making, and alliancing" (2000: 1105). It appears that more empirical studies of dynamic capabilities (e.g., their nature, implementation, and sustainability), and especially review pieces that integrate empirical findings into theories or theory-based design guidelines, would be very useful to organization design researchers focused on investigating designs for the future's more dynamic organizational environments. (For contemporary thinking about the RBV in dynamic environments, see Sirmon et al. 2007.)

Structural *contingency theory* holds that for each level of a contingency (e.g., environmental turbulence) there is a level of a structural attribute (e.g., formalization), which produces the highest performance. More generally, contingency theory explains the nature of an organization as an aggregate of organizational attributes, each of which satisfies one of the various contingencies with which the organization is confronted. Because organizational environments are complex (e.g., possessing multiple important sectors), there are multiple environmental contingencies for (say) the structure to deal with, making it difficult to create a structure in which the multiple attributes of which (each optimized for a different contingency) are not incongruent with each other. Further, organizational features other than structure, such as culture and routines, must be chosen and maintained with respect to internal and external contingencies, creating the possibility of incongruities among the organizational attributes associated with the various organizational features. It seems that contingency theory would be viewed as broader in scope by organization theorists and of more practical value to organization design researchers and practitioners if contingency theorists elaborated the theory to deal with the possible incongruities arising from the existence of (1) multiple contingencies for a specific design feature and/or (2) multiple design features, the optimal attributes of which may be incongruent (see for example Drazin and Van de Ven's (1985) "systems approach" to structural contingency theory). Similarly, the theory would be of additional value if it dealt with the fact that an organization's attributes must change as the organization's environment changes (see, for example, Perez-Nordtvedt et al. 2008, and Siggelkow 2001).

It seems important to call attention to the fact that many organization theorists contribute to *process theories* that are closely related to the organization theories just discussed. Examples are organizational learning, organizational information processing, organizational decision making, and organizational change. Because studies in these areas examine the antecedents and effectiveness of these organizational processes, where effectiveness is generally assessed in terms of variables that contribute to organizational performance and survival (i.e., organizational effectiveness variables), it can be argued that work in these four areas is encompassed within organization theory. Thus, in the next section, when I mention theory or organization theory, I mean to include both prominent organization theories such as those described above and also organizational process theories.

Having seen that numerous opportunities exist for extending current organization theories to be more valid in current and future organizational environments and to be more useful to organization design research and practice, let us examine how the contributors to this volume have used organization theories and the organization theory literature.

Uses of Theory by Contributors to This Volume

I turn now to the use and non-use of theories in the chapters of this volume. My comments are intended not to be evaluative or even descriptive of the chapters. Rather, I intend to use the chapters to suggest – when applicable – the variety of ways in which prominent organization theories (in both their established versions and their currently evolving extensions) can be made more useful in organization design research and practice, and also to suggest how and when they need not be used in these endeavors. After noting the title and authorship of each chapter, I include the authors' abstract and then my comments regarding the chapter's use of theory.

Part I. Towards New Organizational Forms

Blade.Org: A Collaborative Community of Firms

Charles C. Snow, Doreen R. Strauss, and Christopher Lettl

Abstract: The purpose of our chapter is to analyze Blade.org, a community of firms focused on the development and adoption of open blade server platforms (an innovative computer server technology). Founded in early 2006, Blade.org is a successful community of more than 100 member firms which engage in various forms of collaboration to develop innovative products and services and to extend their market reach. Blade.org is a community of firms, but it was purposefully designed to emulate many of the core features and processes of a community of individuals. We argue that a community of firms is a new organizational form, one that increasingly will be found in situations where continuous innovation is a strategic objective.

Commentary: Blade.org possesses certain features that distinguish it to some extent from other types of virtual organizations to which it is related, e.g., user groups and research consortia. Its primary claim to being a new organizational form, however, is that it is a virtual organization composed not of a population of organizations, as are most user groups and research consortia, but of a community of organizations. In their ethnographic study of Blade.org the authors did not seem to draw on any organization theory. On the other hand, their knowledge of the organization theory literature enabled them to recognize that Blade.org was a unique organizational form. (Blade.org, user groups, and research consortia are examples of the types of network organizations that were not within the focus of organizational theorists during the fruitful theory-generation era to which Davis and Marquis (2005) and Dunbar and Starbuck (2004) refer.)

Network-Level Task and the Design of Whole Networks: Is There a Relationship?

Patrick Kenis, Keith G. Provan, and Peter M. Kruyen

Abstract: This paper explores a key argument of structural contingency theory in the context of whole networks of organizations. Specifically, we examine the relationship between the task such networks perform and their design, which we operationalize as one of three forms of governance. Based on an extensive review of the literature on whole networks, we conclude that there is no clear relationship between network-level task and network design. We offer a number of explanations that might account for this finding. Our conclusions help to advance theory and practice on the design of whole networks of organizations.

Commentary: One way that organization design researchers can draw on existing organization theories to benefit organization design practice is to determine whether the theories are valid for organizational forms or organizational environments for which the theories have not previously been evaluated. If the theory is validated for these situations, then organization design guidelines can be derived from the theory. This chapter is a good example of this use of organization theories – in this case, network theory. Structural contingency theory was tested and found to be non-predictive of governance structures for the relatively unstudied combination of organizational task and network governance structure. The authors did find some evidence suggesting that structural contingency theory might be predictive for whole network governance structures when the contingency is the network's size. The study reported is especially interesting because it examines whether a well-known contingent relationship, the effect of task on structure at the level of the individual organization, also holds at the level of the whole network organization. The results indicated that the effect does not hold at this higher level of analysis.

Part II. Dynamics of Adaptation and Change

Organizational Trade-Offs and the Dynamics of Adaptation in Permeable Structures

Stephan Billinger and Nils Stieglitz

Abstract: Organization design has a critical impact on how firms adapt to the business environment. In our case study, we show how organization design increases a firm's ability to sense and seize business opportunities by making its organizational boundaries more permeable. Our findings reinforce and substantiate prior work on organization design and organizational adaptation. They also suggest how insights from organization design theory may help better understand the dynamic capabilities of firms. We find that disintegration and the creation of a permeable corporate structure requires decision-makers to consider four organizational trade-offs: specialization, interdependencies, delegation, and incentives. We discuss how these organizational trade-offs provide a useful complementary perspective to the dynamic capability approach by highlighting the structural properties that shape organizational adaptation across time.

Commentary: Although their data collection process was not shaped with respect to a specific prominent organization theory, the authors explicitly analyzed their qualitatively collected data using four "theoretical concepts that are well supported in various streams of literature," i.e., specialization, interdependence, delegation, and incentives. Aided by, or relying on, these literature-supplied lenses, the authors insightfully identify the evolution of several formal adaptation-managing routines (i.e., dynamic capabilities) which were enabled by the organization's intentional redesign of its structure and which were described by the organization's managers as being very effective for managing organizational adaptations. Thus these organizational design researchers benefitted from their knowledge and use of the organization theory literature (especially the four theoretical concepts mentioned above and also a fifth – dynamic capabilities), but they did not draw on organization theories. It is unclear whether drawing on organization theories would have contributed to a deeper understanding of what they observed during their data collection or whether doing so would have hindered the scope of their insights. It may be that the important lesson to be learned is that organization theory concepts, as contrasted with the theories themselves, might be useful in the analysis of qualitative data and might be less cognitively constraining than would be the use of a particular organization theory in analyzing such data.

Unpacking Dynamic Capability: A Design Perspective

Deborah E. M. Mulders and A. Georges L. Romme

Abstract: This chapter reviews the dynamic capability literature to explore relationships between definition, operationalization, and measurement of dynamic capability. Subsequently, we develop a design-oriented approach toward dynamic capabilities that distinguishes between design rules, recurrent patterns of behavior, operating routines and processes, market and competitive conditions, and performance. This framework serves to develop a number of propositions for further research. As such, we integrate the literature on dynamic capabilities that primarily draws on economics, with a design-oriented approach.

Commentary: Through the creative and careful use of ideas and findings from the dynamic capabilities, organization design, and organizational change literatures, the authors generate propositions which they modestly describe as a "preliminary set of causal claims." While the propositions set forth in the chapter will likely influence future empirical works contributing to theory about dynamic capabilities, the soundness of the propositions also serve as evidence that theory building efforts need not draw on – or even manifest knowledge of – formal organization theories. (The authors may well be knowledgeable about all of the prominent organization theories – that they apparently did not need such formalized knowledge in order to contribute to the development of new theory is the point to be recognized.) While reflection might lead to the discovery of ways that the authors might have used formal theories in their work, the creativity and yet thoroughness of the work suggests that had the authors used organization theories to guide their work they might have adversely constrained their thinking.

Predicting Organizational Reconfiguration

Timothy N. Carroll and Samina Karim

Abstract: This chapter addresses the issue of structural change within for-profit organizations, both as adaptation to changing markets and as purposeful experimentation to search for new opportunities, and builds upon the "reconfiguration" construct. In the areas of strategy, evolutionary economics, and organization theory, there are conflicting theories that either predict structural change or discuss obstacles to change. Our aim is to highlight relevant theoretical rationales for why and when organizations would, or would not, be expected to undertake structural reconfiguration. We conclude with remarks on how these literatures, together, inform our understanding of reconfiguration and organization design and provide insights for practitioners.

Commentary: This chapter contains and builds on a wide-ranging and informative review of literatures reporting on the circumstances under which organizations do, and do not, reconfigure themselves. The authors use this review to develop and frame conclusions about these circumstances. It seems likely that some readers will take the next step and fine tune the conclusions as propositions, thus moving along on the path toward theory development. In other instances, the authors articulate questions unanswerable from current literatures, thus identifying areas for further work. Rather than employing organization theories per se to draw their conclusions, the authors employed organization theories to guide and present their literature review. The range and size of the literatures on which they drew, and for which they provide coherent reviews, indicates that this use of organization theories served them well.

Thus in this chapter and the preceding two chapters we see two different strategies with respect to the use of theory in theory building. On the one hand, the use of theory templates can restrain creative insights and, in a work intended to generate new theory, might be advantageously avoided. On the other hand, the use of the templates can add efficiency to literature reviews and can reduce the chance of being overwhelmed when attempting to draw conclusions about what existing literature has to offer. In either case concepts and ideas associated with prominent theories were important building blocks in the development of theories or the presentation of conclusions that precede theory development.

Embedding Virtuality into Organization Design Theory: Virtuality and Its Information Processing Consequences

Kent Wickstrøm Jensen, Dorthe Døjbak Håkonsson, Richard M. Burton, and Børge Obel

Abstract: What is virtuality in organization design? In this chapter we argue for the importance of understanding the nature and effect of the characteristics of virtual organizations, rather than simply focusing on how these characteristics are different from co-located organizations. Through a review of literature relating to virtual organizations we identify two dimensions: locational and relational differentiation which capture the nature of virtual organizations well. We anchor theoretically these dimensions to organization design and information processing theory. This enables us to identify their effects and consequences for coordination in information processing terms. We thereby not only integrate theory of virtual organization into extant theory of organization design but more importantly, we also demonstrate how increasing virtuality essentially imposes an information processing dilemma for organizations: locational differentiation reduces the information processing capacity, while relational differentiation increases the information processing requirements. We discuss the managerial as well as the theoretical implications of these findings.

Commentary: In my view, the principal contribution of these authors is their portrayal of the organization's information processing requirements as an environmental contingency that must be addressed with information processing routines and structures that provide for the organization's information processing capacity. To do this well, the authors drew on information processing theory and information systems design theory. This chapter is a good example of where organization design researchers used a formal theory (albeit one not prominent in the organization theory literature) in conjunction with a version of contingency theory to make a contribution to the body of organization design research – in the form of a heretofore unarticulated organization design dilemma (that relational differentiation increases information processing requirements and locational differentiation reduces information processing capacity).

Part III. Fit and Performance

Learning-Before-Doing and Learning-in-Action: Bridging the Gap Between Innovation Adoption, Implementation and Performance

Eitan Naveh, Ofer Meilich, and Alfred Marcus

Abstract: Implementation links purpose to outcome. Our model of implementation effectiveness centers on learning – learning-before-doing (preparation) and learning-in-action (adaptation and change catalysis). We explain both the degree of implementation and its impact on various measures of performance (subjective and objective) and test our proposed model on a large, multi-industry sample in the context of implementing the ISO 9000 quality standard. We find that learning-before-doing, an important means for bridging the adoption-implementation gap, is a necessary but not sufficient condition for realizing the benefits of a planned change. To fully bridge the implementation-performance gap, both aspects of learning-in-action – adaptation and change catalysis – must accompany implementation.

Commentary: This chapter is a good example of how organization theory (and organization design theory – the findings can easily be framed as design guidelines) is developed. That is, the authors draw on organization learning theory and organization change theory to develop propositions concerning organization change routines that would contribute to performance and test the propositions in a richly described field setting.

Underfits Versus Overfits in the Contingency Theory of Organizational Design: Asymmetric Effects of Misfits on Performance

Peter Klaas and Lex Donaldson

Abstract: The contingency theory approach to organizational design traditionally treats underfit as producing equal performance loss as overfit, so that the effects of these misfits on performance are symmetrical. Recently, an asymmetric view has been proposed, in which underfit produces lower performance than overfit. The paper analyzes these views. The effects of underfits and overfits on benefits and costs are distinguished. The differential effects on organizational performance of underfit and overfit are to be understood by their effects on benefit and costs. Implications are drawn for organization theory and design. For future empirical research, it is specified how to correctly identify differential performance effects of underfits and overfits. In a managerial design perspective, underfit is liable to occur in growing organizations and to rob them of some of their potential growth. While underfit will lead to an acute condition, overfit will be more chronic.

Commentary: Arguably mistaken beliefs in scientific literatures are often challenged with empirical studies. In contrast, the authors of this chapter draw on their deep understanding of contingency theory, on the contingency theory literature, and on their previous analyses, to extend contingency theory by examining the asymmetric effects of misfit on organizational performance. In so doing, they disabuse us of the frequently encountered (mis)assumption that underfit produces the same performance effect as does overfit, and they describe the implications for organization design researchers and organization designers. Thus, by extending the established version of an organization theory, the authors contribute to organization design research and practice.

Based on this small but rich convenience sample of organization design research studies, what might we infer about how current organization theory is serving the organization design community? Below I offer some possible answers to this question.

Concluding Observations

At the beginning of this introduction we encountered the idea that several prominent organization theories were of limited usefulness for understanding organizations in current organizational environments. Our subsequent examination of these theories suggested that this idea had validity, but also that the theories did serve as foundations for extensions of the theories, extensions that could be useful for understanding organizational performance and survival in current and future organizational environments and for the conduct of organization design research and practice.

From the review of how the authors of the chapters in this volume used prominent organization theories, it seems to me that we can draw two interesting and useful inferences.

One is that currently prominent theories can be tested for their validity in current or evolving organizational environments. In this way, and as the foundation for outright extensions, these theories can be made to contribute to organization design research and practice. The theories can also serve to guide and present literature reviews and the conclusions that result from the reviews, conclusions that often serve as the bases for propositions to be examined for their possible contributions to organization design theory.

A second possible inference from the review of the chapters is that, as contrasted with the theories themselves, the ideas and concepts associated with the organization theory literature can serve as building blocks in the development of new theory. The ideas and concepts can also serve as cues and lenses for recognizing new organizational phenomena and forms and for interpreting the results of ethnographic and qualitative studies.

Together these inferences suggest to me that, rather than ignoring the theories that were developed in and for earlier organizational environments, we should view the theories as phases in the evolution of organization theory and as contexts for better understanding concepts and ideas that are useful and perhaps necessary for developing organization design theory for the future.

References

Aldrich HE, Ruef M (2006) Organizations evolving. London: Sage.

Barnett WP, Carroll GR (1995) Modeling internal organizational change. Annual Review of Sociology 21: 217–236.

Child J (1997) Strategic choice in the analysis of action, structure, organizations and environment: Retrospect and prospect. Organization Studies 18: 43–77.

Daft RL (2007) Organization theory and design. Boston: Thompson South-Western.

Daft RL, Lewin AY (1990) Can organizational studies break out of the normal science straight-jacket? Organization Science 1 (1): 1–9.

Daft RL, Lewin AY (1993) Where are the theories for the "new" organizational forms? An editorial essay. Organization Science 4 (4): i–vi.

Davis GG, Marquis C (2005) Prospects for organization theory in the early twenty-first century: Institutional fields and their mechanisms. Organization Science 16 (4): 332–343.

Dobrey SD, Kim TY, Carroll GR (2002) The evolution of organizational niches: U.S. automobile manufacturers, 1885–1981. Administrative Science Quarterly 47 (2): 233–264.

Drazin R, Van de Ven AH (1985) Alternative forms of fit in contingency theory. Administrative Science Quarterly 30 (4): 514–539.

Dunbar RLM, Starbuck WH (2004) Call for papers. Organization Science 15 (14): 375–497.

Eisenhardt KM, Martin JA (2000) Dynamic capabilities: What are they? Strategic Management Journal 21: 1105–1106.

Hambrick D (2007) Upper echelons theory: An update. The Academy of Management Review 32: 334–343.

Haunschild P, Chandler D (2008) Institutional-level learning: Learning as a source of institutional change. In: Greenwood R, Oliver C, Sahlin-Andersson K, Suddaby R (eds), The Sage handbook of organizational institutionalism. California: Sage.

Huber GP (2004) The necessary nature of future firms: Attributes of survivors in a changing world. California: Sage Publications.

Kerlinger FN (1986) Foundations of behavioral research. New York: Holt, Rinehart and Winston.

Perez-Nordtvedt L, Payne GT, Short JC, Kedia BL (2008) An entrainment-based model of temporal organization fit, misfit, and performance. Organization Science 19 (5): 785–801.

Provan KG, Sebastian JG (1998) Networks within networks: Service link overlap, organizational cliques, and network effectiveness. Academy of Management Journal 41: 453–463.

Rosenkopf L, Padula G (2008) Investigating the microstructure of network evolution: Alliance formation in the mobile communications industry. Organization Science 19 (5): 669–687.

Scott WR, Davis GF (2007) Organizations and organizing: Rational, natural, and open system perspectives. Upper Saddle River: Pearson Prentice Hall.

Siggelkow N (2001) Change in the presence of fit: The rise, the fall, and the renaissance of Liz Claiborne. Academy of Management Journal 44 (4): 838–857.

Sirmon DG, Hitt MA, Ireland RD (2007) Managing firm resources in dynamic environments to create value: Looking inside the black box. Academy of Management Review 32 (1): 273–292.

Soda G, Usai A, Zaheer A (2004) Network memory: The influence of past and current networks on performance. Academy of Management Journal 47: 893–906.

Teece DJ, Pisano G, Shuen A (1997) Dynamic capabilities and strategic management. Strategic Management Journal 18: 509–533.

Uzzi B (1996) The sources and consequences of embeddedness for the economic performance of organizations: The network effect. American Sociological Review 61: 674–698.

Huber GP (forthcoming) Organizations: Theory, design, future. In: Zedeck S (eds), Handbook of Industrial/Organizational Psychology, Vol. 1. Washington: American Psychological Association.

Walsh JP, Meyer AD, Schoonhoven CB (2006) A future for organization theory: Living in and living with changing organizations. Organization Science 17 (5): 657–671.

Part I
Toward New Organizational Forms

Chapter 1
Blade.Org: A Collaborative Community of Firms

Charles C. Snow, Doreen R. Strauss, and Christopher Lettl

Abstract The purpose of our chapter is to analyze Blade.org, a community of firms focused on the development and adoption of open blade server platforms (an innovative computer server technology). Founded in early 2006, Blade.org is a successful community of more than 100 member firms which engage in various forms of collaboration to develop innovative products and services and to extend their market reach. Blade.org is a community of firms, but it was purposefully designed to emulate many of the core features and processes of a community of individuals. We argue that a community of firms is a new organizational form, one that increasingly will be found in situations where continuous innovation is a strategic objective.

Keywords Community of firms · Collaborative community · Collaborative innovation networks · Open source innovation · New organizational forms

1.1 Introduction

More than seven decades ago, economist Joseph Schumpeter (1934) posited that innovation was the main driver of economic development. Today, the idea that a firm's long-term competitiveness hinges on its ability to innovate is a widely accepted fact (Baumol et al. 2007; Hult et al. 2004; Zhonqi et al. 2004). Despite its importance to firm growth and success, innovation is not an easy task to accomplish in the typical firm. Indeed, one survey found that chief executive officers believe their firms utilize only 15–25% of their innovation capacity (Käser and Miles 2002). Various approaches to improving a firm's ability to innovate have been developed, including cross-functional business teams, internal venture capital processes, creating or acquiring new business units, spinning off new ventures, and forming alliances with or investing in partner firms. All of these approaches, however, tend to produce periodic and incremental innovation that is mostly limited to the firm's

---

C.C. Snow (✉)
Department of Management and Organization, Smeal College of Business, The Pennsylvania State University, University Park, PA 16802, USA
e-mail: csnow@psu.edu

A. Bøllingtoft et al. (eds.), *New Approaches to Organization Design*, Information and Organization Design Series 8, DOI 10.1007/978-1-4419-0627-4_1,

existing businesses – not the continuous and radical innovation that is increasingly required by today's global economy.

In recent years, a new philosophy of innovation has emerged. Referred to as "open innovation" (Chesbrough 2006, 2007), this philosophy rests on the belief that a firm's innovation success comes from opening up its innovation processes to external sources of knowledge and creativity. Chesbrough (2006: 24) describes open innovation as a "...paradigm that assumes that firms can and should use external as well as internal ideas, and internal and external paths to market...." Open innovation processes are becoming increasingly collaborative and democratic (von Hippel 2005). Enabled by advances in information and communication technologies, individuals and groups outside of established firms are nowadays not only able to develop new products themselves but also are willing and able to share their knowledge, experiences, and innovative concepts with a community of peers at relatively low cost (von Hippel 2005, 2007). Those individuals or groups are often "lead users" of new products and technologies, and lead users have been shown to be an important source of innovation in many industries (Lilien et al. 2002; von Hippel 1988, 2001).

The newest form of open innovation has been called a "collaborative community of firms" (Miles et al. 2005). The community-of-firms' organizational model differs from previous approaches to open innovation in three main ways. First, unlike most communities described in the organization sciences literature (e.g., von Hippel 2005; Wenger 1998), community members are firms rather than individuals. Second, communities of firms have an economic as well as a technical purpose. They have been established to both create and commercialize new products and services. Third, most open innovation communities are focused on the improvement of a particular subject or field such as the Linux operating system (Lee and Cole 2003). A community of firms, by contrast, is organized as an arena in which firms can collaborate with one another to create their own unique solutions for which they retain ownership rights. Launched in early 2006, Blade.org is now one of the largest and most influential communities of firms to emerge from the open innovation movement. In this chapter, we describe the evolution of Blade.org and highlight its collaborative features and processes. We also explain why we believe that Blade.org represents a viable new organizational form, and we discuss implications for organization design theory and practice.

1.2 Organizing for Collaborative Innovation

Due to the importance of new products, services, and technologies to firms' growth, innovation in many companies has traditionally been a closely guarded process. Companies relied on their internal R&D capability to invent and commercialize products and services on their own (Gassmann 2006). Although market and other external information may have been gathered, analyzed, and distributed, outsiders frequently played a passive and limited role in the innovation process.

Today, however, it is evident that a paradigm shift is underway. The shift toward open innovation has been prompted by several factors (Chesbrough 2007). First, the competitive pressure to innovate is increasing in many industries due to rapid technological advancements. Second, the cost of developing new technologies and products is increasing due to greater product complexity and proliferation. Third, product life cycles are shortening in many industries, giving firms less time to amortize their investments in new product development. To solve the problem of developing new and more complex technologies and products at an even faster pace, firms need to leverage their and other firms' resources extensively. This can be achieved by exploiting ideas and technologies developed outside the firm's boundaries as well as by allowing unused ideas and technologies to flow to the outside (Chesbrough 2006; Huston and Sakkab 2006).

Open innovation was originally fueled by communities of individuals, a social innovation that is a major cultural phenomenon (von Hippel 2001, 2007; von Hippel and von Krogh 2003). Communities of individuals are social networks in which professionals and/or hobbyists voluntarily exchange new ideas, knowledge, and experiences about a particular topic or field of interest, thus creating an intellectual "commons" (Harhoff et al. 2003; von Hippel 2001, 2007; Wenger 1998).

1.2.1 Scope and Types of Communities of Individuals

Communities of individuals have appeared in a wide variety of fields and have assumed many different forms. One well-known form is the community of practice which refers to the process of social learning that occurs when individuals who have a common interest in a particular topic or field collaborate over an extended period of time to share ideas, find solutions, and build prototypes. Communities of practice have emerged both within and across firms (Lave and Wenger 1991; Wenger 1998).

A prominent community of practice which has gained attention from researchers and practitioners alike is open source software where communities such as those associated with the Apache web server and the Linux operating system have developed not only software solutions with superior market value but also new business models and successful new firms. Also, products in both high-tech industries (e.g., medical equipment, biotechnology, nanotechnology) and low-tech industries (e.g., sports equipment) increasingly are being developed by communities of professionals either independently or in collaboration with incumbent firms. The diverse forms of communities of practice can be classified into a typology whose main dimensions are organizational scope and degree of creativity.

With respect to organizational scope, three archetypes can be distinguished. First, communities can be primarily collectives of employees within a firm. This type of community enhances the learning processes by which an organization comes to "know what it knows" and thus is effective and valuable as an organization (Brown and Duguid 1991, 2000, 2001; Wenger 1998). For example, Siemens has established more than 20 internal communities of practice spanning organizational units as the anchor of its knowledge management system (Zboralski et al. 2006). The scope of this type of community of individuals is restricted to the knowledge-generating potential of those employees within a single firm. Second, communities can form completely outside the boundaries of established firms (von Hippel 2001, 2005, 2007). For example, Zeroprestige.com is a community of professional and hobbyist kite-flyers who develop next-generation kite designs by using aerodynamic modeling and design toolkits stored on the community's web site. Third, hybrid forms exist in which communities of individuals collaborate with one or more incumbent firms (Jeppesen and Frederiksen 2006; Sawhney and Prandelli 2000). For example, the AO Foundation (www.aofoundation.org) is a network of leading orthopedic surgeons who work with industrial firms in the areas of research, development, licensing, and education. Hybrid forms also occur when communities of individuals emerge around the brands of producer firms. Examples of such brand communities are LEGO user communities (Antorini 2005) or the STATA community of users of sophisticated statistical software (Harhoff and Mayrhofer 2007). An interesting manifestation of a hybrid community is when a "mediating" (Stabell and Fjeldstad 1998; Thompson 1967) firm such as Innocentive.com (www.innocentive.com) creates an online infrastructure where established firms can post R&D challenges – those that they are unable to solve by themselves – to a global community of scientists (Lakhani et al. 2007).

A second major dimension to classify communities of individuals is the extent to which the community demonstrates creative processes such as idea generation, technical problem solving, and prototype development. While there are many communities which have the explicit purpose of developing new solutions or techniques, there are also many communities whose primary focus is the mere exchange of experiences and information (discussion forums). For the remainder of this section, we focus our discussion on communities of individuals that explicitly seek to develop novel concepts and innovative solutions in their respective fields.

1.2.2 Key Characteristics of Communities of Individuals

Despite the large variety of topics/fields and different community types, communities of individuals share several characteristics. Foremost, they are a forum where individuals with different backgrounds but a common interest in a specific topic or field voluntarily share their experiences and exchange diverse information and knowledge, both technical and market-related. Such knowledge sharing allows a group of individuals to collectively develop new knowledge, post and solve particular problems, pursue innovative ideas, and develop new applications for existing or new technologies (Baldwin et al. 2006; Franke and Shah 2003; Jeppesen and Frederiksen 2006; Shah 2005; von Hippel 2005, 2007; von Krogh and von Hippel 2006; von Krogh et al. 2003; Wenger 1998). A key characteristic of communities of individuals, therefore, is that proprietary knowledge, novel ideas, and solutions are freely revealed among members, and any intellectual property generated within the community belongs to the entire community (Harhoff et al. 2003; Sawhney and Prandelli 2000; von Hippel 2001, 2005, 2007).

From an economic perspective, the behavior of individuals within a community may seem to be irrational. Lerner and Tirole (2002), for example, pose the issue this way: Why do individuals dedicate their talent, creativity, and time to the development of a public good? Research is only just beginning to develop answers to this question. In economic terms, the preliminary answer is that the process of collectively developing a public good can also produce private goods (von Hippel and von Krogh 2003). Private goods are a mixture of intrinsic and extrinsic rewards. Intrinsic rewards include the enjoyment of creating something new, the stimulation of an intellectual challenge, and feelings of reciprocity, identity, and solidarity with a group or community. Extrinsic rewards include an enhanced reputation, the signalling of technical excellence to potential employers and venture capitalists, and various other benefits derived from better solutions and their faster diffusion (Ryan and Deci 2000).

Both intrinsic and extrinsic motivations at least partly explain why communities typically emerge spontaneously through informal contacts, outside the normal planning, incentive, and control systems used in formal organizations. Communities are managed and sustained voluntarily rather than hierarchically. By making contributions to a community, individuals develop a shared identity, trust, and collaborative values and norms regarding good community citizenship (Raymond 1999; Wenger 1998). Within communities of individuals, shared identity, collaborative values, and trust serve as substitutes for contractual arrangements and other formal control mechanisms (Bagozzi and Dholakia 2006; Brown and Duguid 2001; von Hippel and von Krogh 2003).

1.2.3 Mechanisms That Facilitate Innovation Within Communities of Individuals

As a means of creating innovations, communities of individuals are an ever-expanding source of diverse technical knowledge that can be used to develop new products and services. While community members have a common interest in the topic of the community, each community member offers a distinct knowledge set based on his or her professional background and experience. Communities of individuals, therefore, are able to produce valuable knowledge that often exceeds that found within the R&D labs of a producer firm or within a group of individuals.

The LEGO Mindstorms user community is a case in point (http://mindstorms.lego.com/community/default.aspx). Mindstorms is a LEGO brick robot which has a computer "brain," stepper motor for movement, and different types of sensors (e.g., light, touch, temperature). LEGO introduced Mindstorms in 1998, and a community quickly formed around it to advance the robot technologically. LEGO enthusiasts who are at the same time software hackers, sensory technology experts from NASA, and university professors in the area of robotics collaborate to improve the robot's functionalities. By applying community problem solving to the field of robotics, the LEGO Mindstorms community has outdistanced the product extensions developed in LEGO's central R&D labs in Denmark (Koerner 2006).

Knowledge diversity within communities allows individuals to break out of established solution trajectories and problem-solving paradigms. When a specific problem is posted, community members self-select to solve it based on relevant knowledge that they have in stock (Kogut and Metiu 2001; von Hippel 2001, 2007). The self-selection process ensures that a given problem is viewed from different but relevant angles. In this respect, communities facilitate self-evolving analogical reasoning as community members contribute knowledge from diverse base domains to the target domain (i.e., the focus of the community). Analogical reasoning has been shown to be a driver of creative thought leading to novel solutions (Boden 1994; Dahl and Moreau 2002; Ebadi and Utterback 1984; Hargadon 2003; Lakhani et al. 2007; Ward 2004). Furthermore, sharing knowledge from diverse disciplines enables the development of new combinations of existing but unconnected knowledge sets. Due to the high amount of knowledge diversity, communities of individuals often can reach a higher "absorptive capacity" (Cohen and Levinthal 1990) than established firms.

Communities of individuals also provide a setting for the collective assessment of solutions. Novel ideas that are posted in the community are reviewed by peers. Similar to review processes in scientific communities, the originator of an idea receives valuable suggestions and comments on how to improve the idea. The more reviewers of an idea or a solution, the more likely that technical problems and adoption barriers can be identified. Thus, communities enable solution refinement (Jeppesen and Molin 2003; Pruegl and Schreier 2006; Raymond 1999). Furthermore, user communities can serve as a test market for proposed ideas and concepts. As each idea receives feedback from the community, members collaboratively select those ideas which are the most promising.

In summary, communities of individuals incorporate mechanisms that have been identified as crucial to the innovation process (see Table 1.1). One way that a commercial firm can benefit from such innovation capability is to participate in relevant communities. Another newer approach is to build a community of firms, thereby elevating the community model of organizing from the level of the individual to the firm level. The overall challenge in designing such a community is to purposefully create collaborative processes that emulate what has evolved spontaneously in communities of individuals. Specifically, designing such a community involves

Table 1.1 Communities of individuals: characteristics and mechanisms for facilitating innovation

Community characteristics	Innovation mechanisms
Spontaneous emergence based on a shared interest	Application of diverse knowledge and experience sets
Voluntary membership	Collective development of solutions
Open sharing of knowledge and information	Collective assessment of solutions
Trust based on collaborative values and norms	Test market for proposed ideas and concepts (collaborative filtering)
Community ownership of intellectual property	Common development of technical standards

the (a) adoption of collaborative values and open innovation principles, (b) attraction of a large number of actively contributing firms, (c) development of processes that facilitate all stages of innovation from invention to commercialization, and (d) use of a governance structure that is voluntary rather than hierarchical. To illustrate the community-of-firms model, we describe Blade.org – its origin and purpose, membership and governance structure, and collaborative innovation processes.

1.3 Blade.Org: The Building of a Collaborative Community

Blade.org is a collaborative community of firms focused on the development and adoption of open blade server platforms, an innovative computer server technology developed by IBM. Blade.org was established in early 2006 by IBM, Intel, and six other founding firms to increase the number of blade platform "solutions" available for customers and to accelerate the process of bringing them to market. From the original eight founding companies, Blade.org has grown to a community of more than 100 member firms including leading blade hardware and software providers, developers, distribution partners, and end users from several countries (though most of Blade.org's member firms are in the United States). Our research team has been studying Blade.org since May 2007.[1] Members of the team have conducted multiple interviews with leaders in Blade.org's principal office and with various executives and technical specialists in five member firms, including two of the founding firms and one international member firm. We have also attended, and gathered data at, two All-Member Meetings of Blade.org which are held three times a year. Lastly, in May 2008 we conducted an online survey of Blade.org member firms that focused specifically on community-of-firms issues and processes. We remain in contact with the executives in Blade.org's principal office in order to monitor and analyze organizational developments as they occur.[2] Our research methods are summarized in Table 1.2.

1.3.1 Origin and Purpose

The origin of Blade.org can be traced to August 2004 when IBM announced that it was opening the specifications to its BladeCenter server chassis (Clabby Analytics 2007). IBM stated that its goal was to build a developer community that would

---

[1]The members of our international research team are Raymond Miles (University of California, Berkeley), Charles Snow (The Pennsylvania State University), Grant Miles (University of North Texas), Kirsimarja Blomqvist (Lappeenranta University of Technology, Finland), Doreen Strauss (The Pennsylvania State University), Christopher Lettl (University of Aarhus, Denmark), and Øystein Fjeldstad (Norwegian School of Management). This chapter is based heavily on Doreen R. Strauss, Blade.org: A New Beast in the Jungle? Honors Thesis, The Pennsylvania State University, 2008.

[2]For additional research reports on Blade.org, see Miles et al. (forthcoming), and Snow et al. (2009).

Table 1.2 Four-phase data collection process

Phase	Data source	Topic focus	Unit of analysis	Number of firms	Time of data collection
1	Familiarization interviews, web-based archival searches, and attendance at an All-Member Meeting	Community objectives, founding firms, member firms, technology, industry setting	Firm and community	Two (as well as interviews at an All-Member Meeting)	February–April 2007
2	Semi-structured interviews, Blade.org web site analysis, and Blade.org by-laws	Values and experiences, crucial organizational design features, collaborative innovation processes among member firms, and potential benefits to founding and member firms	Community	Two (as well as interviews at Blade.org's principal office)	October 2007–January 2008
3	Online survey	Reasons to join Blade.org, collaborative innovation processes among member firms, required investments, derived benefits, etc.	Member firm	46	April–May 2008
4	Follow-up interview	Points of clarification, emerging community issues	Community	One (principal office executive)	September 2008

focus on expanding the number of solutions that could be made available from its promising blade architecture. IBM also noted that it could not drive all innovation on blade applications itself; it expected its partners to play a major role in creating future blade-based solutions. In February 2006, IBM announced the formation of an independent organization (Blade.org) that would serve as the facilitator for a community of blade developers, and it invited vendor and user firms to provide feedback and develop products specifically for BladeCenter. A timeline showing the major milestones in Blade.org's development is shown in Fig. 1.1.

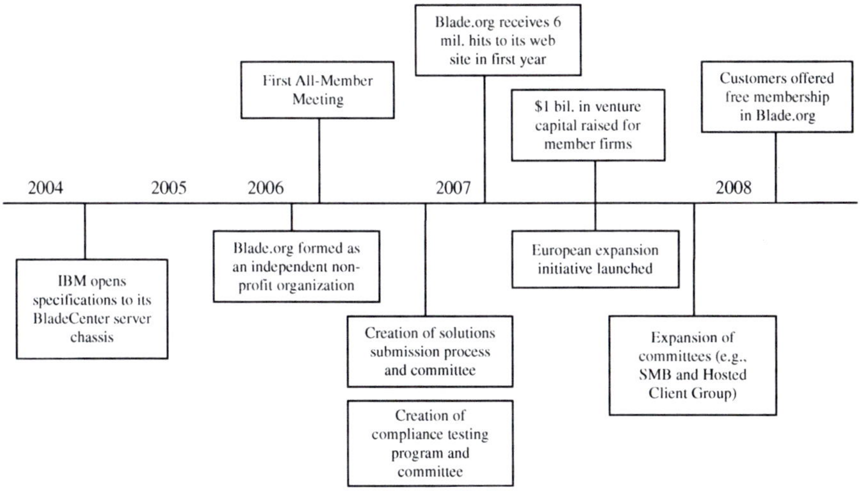

Fig. 1.1 Milestones in Blade.org's Development

The overall economic purpose of Blade.org is to foster and accelerate the growth of solutions based on the blade processor technology. The specific purposes for which Blade.org is organized include enabling the ongoing development of blade-based solutions, helping to bring solutions to the market in a timely fashion, increasing the adoption and number of solutions in both existing and new markets, and increasing customer confidence in blade-based solutions. The Blade.org community undertakes a wide variety of activities to achieve these purposes, including the provision of guidelines to member firms for designing their solutions, developing independent compliance testing procedures that member firms may use, hosting industry-wide SolutionFests and other marketing events, educating the marketplace on blade platform solutions, and incorporating member concerns and preferences into strategic initiatives that expand and improve the community.

1.3.2 Membership and Governance Structure

Blade.org has three membership categories: Governing Members, Sponsoring Members, and General Members. Governing Member firms, each of which has a representative who sits on Blade.org's board of directors, are limited by the organization's bylaws to 11 in number and include the original eight founding firms (Brocade, Citrix, IBM, Intel, Network Appliance, Nortel, Novell, and VMWare). Governing Member firms pay annual membership dues (as do all member firms except customers), entitling them to certain rights (www.Blade.org Membership Benefits 2008; Bylaws of Blade.org 2006):

- Opportunities to collaborate with other Blade.org solution providers
- Influence the direction of the blade server market

- Networking opportunities with industry leaders at trade shows and other industry events
- Increased visibility within the marketplace
- Ability to leverage Blade.org's marketing activities including use of the Blade.org logo in promotional literature
- Use of independent compliance testing arranged by Blade.org
- Increased media coverage through access to Blade.org's public relations firm
- Speaking opportunities at Blade.org events
- Free banner advertising on the Blade.org web site and various discounts
- Ability to appoint a member of the Board of Directors
- Eligibility for their employees to serve as chair of a committee or subcommittee and to participate in the activities of any committee or subcommittee
- Influence the agenda of All-Member Meetings.

Of these various rights, the core rights that accrue from membership in Blade.org are opportunities to collaborate with other member firms and eligibility for participation in the work of committees and subcommittees.

Sponsoring Member firms are of three types: (1) They are currently distributing or developing hardware, software, or services offerings for the blade platform; (2) they provide consulting or distribution support for blade-based solutions or products; or (3) they currently use blade platform solutions. Sponsoring Members have the same rights as Governing Members except for the right to appoint a representative to the Board of Directors. Lastly, a firm can become a General Member of Blade.org if it has a legitimate business interest in participating in the community and is willing to publicly support Blade.org and its mission by being listed on the organization's web site and in press releases. A General Member must be approved by a majority vote of the Board of Directors.

In early 2008, Blade.org began to offer free membership to its customers (called "end users"). End user membership benefits include invitations to participate in a variety of technical and marketing activities, an opportunity to join any Blade.org committee or subcommittee, access to a forum where they can voice concerns and suggestions directly to Blade.org vendors, and an opportunity to network with other firms which allows customers to share best practices within the blade community. Overall, such benefits allow customers to influence the direction of the blade market as well as technology development.

Blade.org operates as a "program" under the auspices of the Industry Standards and Technology Organization (ISTO). ISTO, whose parent organization is the well-established Institute of Electrical and Electronics Engineers (IEEE), was started in 1999 as a not-for-profit corporation that offers industry groups (e.g., consortia, special interest groups, alliances, forums, working groups) support for technology and standards development. The IEEE-ISTO serves as an umbrella organization to provide a legal forum for industry groups to operate without the need to incorporate. Programs of the IEEE-ISTO enjoy the legal protections and insurance benefits of operating within an incorporated, fully insured, nonprofit organization. The IEEE-ISTO provides a complete menu of management and operational support, leaving Blade.org's member firms free to focus on the community's mission and activities.

Blade.org has a principal office located in Research Triangle Park, North Carolina (contiguous to the cities of Raleigh, Durham, and Chapel Hill). The principal office houses the strategic leadership of Blade.org, and its executives plan and organize strategic initiatives designed to expand and enrich the community. In addition, Blade.org has nine committees comprised of volunteers from the member firms. These committees are organized by function and include the Technology Committee, Power and Cooling Committee, Compliance and Interoperability Committee, and Marketing Committee (see Appendix for a list and description of all nine committees). The Blade.org volunteer committees perform a dual function for the community of firms: They do work that is useful to the community as a whole, and they serve as a repository of knowledge that member firms can tap into when needed. Thus, the committees are keepers and developers of the community's intellectual commons.

1.3.3 Collaborative Innovation Processes

A computer server is a computer dedicated to running a server application. A server application is a computer program that accepts connections in order to service requests by sending back responses. Examples of server applications include web servers, e-mail servers, database servers, and file servers. Blade servers are ideal for specific purposes such as web hosting and cluster computing. As more processing power, memory, and I/O bandwidth are added to blade servers, they can be used for larger and more diverse workloads. "Blades" are small dart-shaped devices which, when plugged into a rectangular enclosure, perform as full-fledged computer servers. They are called blades because of how small and thin they are. A blade server "solution" is comprised of two hardware components: the blades themselves and the enclosure that holds them. The enclosure housing the blades is configured to fit a customer's data storage and computing needs. A particular enclosure might hold more than a dozen servers, but the overall capacity depends on the functionality built into the enclosure. Blade technology is innovative primarily because of consolidation – the blades are designed to have high density and power in a small space. The main benefits to a company that buys a customized solution are lower fixed costs due to the smaller physical space required to house the equipment, lower energy costs to operate the equipment, and ease of maintenance and data management tasks.

In less than two years after the establishment of Blade.org, member firms developed 60 solutions, an indication of the overall success of the community in creating innovative products and services. Solutions are explored through a variety of knowledge-sharing processes including web site postings, the work of the nine specialized committees, and participation in All-Member Meetings and other community events. Interfirm collaboration occurs both within and outside Blade.org in the sense that member firms collaborate with their customers (end users) in the development of customized solutions, and they collaborate with one another to produce solutions for existing or new customers. On any given innovation project, collaboration can take one or more of four basic forms: (1) *bilateral collaboration* (this type of innovation occurs when a Blade.org member firm collaborates with its customers on a new solution, perhaps using consulting advice from IBM as the inventor of the blade technology); (2) *direct collaboration* (two to four member firms work together on the development of a new solution); (3) *pooled collaboration* (Blade.org member firms supply ideas, information, and experiences to a central database called Bladeuser.org that is accessible by other member firms to pursue innovation projects); and (4) *external collaboration* (a Blade.org member firm works with a non-Blade.org firm on a "one-off" innovation project).

1.4 Organizational Analysis of Blade.Org

Is Blade.org "a new beast in the jungle" – that is, does it represent a new organizational form? Based on four assessment criteria that pertain to the emergence of new organizational forms, we believe that Blade.org is indeed an example of a new way of organizing. First, according to Palmer and Dunford (2002, 2008), one must ask, what is the meaning of "new" in referring to an organizational form? Does new refer to time (the first time this form has appeared) or to context (the first time this form has appeared in a particular setting)? In both senses, the community-of-firms model is a new organizational form. Although several communities of firms exist, few of them have been purposefully designed or have achieved the size and influence of Blade.org. Thus, Blade.org demonstrates the sustainability of a planned community of firms. Moreover, the community-of-firms model was pioneered by firms in the computer industry. Perhaps this should not come as a surprise since the computer industry, along with other knowledge-intensive industries such as biotechnology, has served as an incubator for the development of new organizational forms in recent years (Powell 1996; Powell et al. 1996).

The other three criteria for assessing organizational forms have to do with the defining characteristics of organizations. Organizations are goal-directed, boundary-maintaining, socially constructed systems of human activity (Aldrich 1979). Most newly formed organizations are "reproducer" organizations in that their purposes, capabilities, and activities vary little, if at all, from those of existing organizations (Aldrich and Ruef 2006). "Innovative" organizations, by contrast, are those which vary significantly from existing organizational forms (Ruef 2002). The principal office of Blade.org was set up specifically to serve as the *facilitator* of a community of firms – its purpose is not to direct and control the firms in the community but rather to create the conditions under which those firms can collaborate with one another. Moreover, because it operates as an IEEE-ISTO program, Blade.org itself does not have the goal of making a profit – though, of course, all of the member firms are profit-seeking. Thus, Blade.org's purpose and goals are innovative in that the organization is oriented toward helping its member firms pursue their own commercial and technical objectives.

Blade.org is also a new organizational form because of how its organizational boundaries have been defined. The basic building blocks of the community are firms, so its boundaries are much wider than those of the usual organization. Blade.org's boundaries are also flexible and permeable, and the community has thoughtfully incorporated venture capitalists, customers, and other groups into its activities as it has evolved. Although organizing a community of firms is very challenging, Blade.org quickly achieved a significant amount of coherence and integrity as an organization.

Lastly, Blade.org represents a new type of activity system – what Hunt and Aldrich (1998) refer to as a "competence-extending" innovation. Blade.org operates a set of collaborative innovation processes side by side with processes of cooperation and even competition. Collaborative processes, both among individuals and organizations, are easier to develop if they are kept separate from the other relational modes. Arguably, collaborative capability is also easier to develop if it does not involve commercial objectives. Thus, Blade.org's activity system is new in that this community of firms is pursuing economic ends through the simultaneous use of collaboration, cooperation, and competition. Such an activity system appears to be especially well suited to the pursuit of continuous and radical innovation.

Blade.org's main organizational features are summarized in Table 1.3. As indicated, the purpose of the organization is to create an arena in which the member firms can collaborate with one another to develop new products and services. The

Table 1.3 Organizational features of Blade.org

Purpose	To provide a means for member firms to collaborate with one another to create and commercialize new products (solutions) based on the blade processor technology
Components	Independent firms that have been admitted to the community
	Committees that (a) ensure technical standards and interoperability across solutions (e.g., the Solutions Committee), (b) develop and maintain common knowledge needed by the member firms (e.g., the Power and Cooling Committee), and (c) legitimate and market the community to the outside world (e.g., the Marketing Committee)
	An institutional context that includes legal bylaws, protection for intellectual property developed by the member firms, and an idea bank (Bladeuser.org) and web site that member firms can use to post and/or obtain ideas for the development of new solutions
Coordination and alignment mechanisms	Self-coordination by member firms through direct networking on collaborative innovation projects
	Governance structure built on the facilitative activities of the principal office and administrative services provided by IEEE-ISTO

main difference between the purposes of a community of firms such as Blade.org versus a community of individuals is economic in nature. Blade.org has a commercial as well as a technical purpose, and the community is focused on product innovation not the mere improvement or extension of a specific topic or field of interest.

The key organizational components of Blade.org include its membership of independent firms, the specialized committees that support the various collaborative innovation processes among firms, and the internal institutional context that has been specifically designed to enable the firms to collaborate effectively with one another. This particular set of organizational components cannot be managed by using the traditional planning, incentive, and control systems associated with hierarchical organizations. Instead, Blade.org's governance system is based on facilitation and voluntarism. That is, the member firms essentially self-coordinate by directly contacting those firms they wish to work with, and they provide the representatives who voluntarily serve on committees that perform activities which benefit the community as a whole. The community expands through the strategic initiatives identified and developed by the principal office, and various administrative services are provided to the member firms by IEEE-ISTO, a non-profit organization that helps industry groups to work together.

Although every organizational form has its own particular limitations, Blade.org has yet to experience any significant dysfunctional or exploitive behavior on the part of its member firms. A community-based organization design would appear to be susceptible to free riding and the appropriation of others' ideas, but neither problem has occurred at Blade.org. Our interview data suggest that Blade.org has been able to avoid such problems because of its careful selection of reputable and competent member firms and from its well-conceived organization design that promotes desirable collaborative behaviors.

1.5 Implications for Organization Design Theory and Practice

The Blade.org experience provides several implications for organization design theory and practice. The main implication is that the focus of organization design theory needs to continue to expand beyond the boundaries of the single firm and multi-firm networks (Gulati 2007) to include community-based organization designs (Demil and Lecocq 2006; Miles et al. forthcoming). The overall goal of such theorizing should be the identification and description of designs that enhance collaborative behavior among firms without restricting the ability of the individual firms to continue to compete within their own marketplace. One promising theoretical framework that addresses such interfirm behavior is that of Dyer and Singh (1998) who identify four potential sources of interorganizational competitive advantage: (1) relation-specific assets, (2) knowledge-sharing routines, (3) complementary resources and capabilities, and (4) effective governance. The community-based design of Blade.org reflects all four sources of interorganizational advantage but especially the power of knowledge-sharing routines (the community supports four different types of collaboration) and an effective governance structure (Blade.org's "facilitator" approach has a strong fit with its overall collaborative innovation process).

A second implication concerns the role of community facilitator. A facilitator, such as Blade.org's principal office, does not operate the organization's core activity system itself but rather creates and maintains the conditions under which the organization can survive and grow. For example, early on Blade.org signaled its member firms that its primary role was to advance the entire community's interests (it operates as a non-profit organization but helps its member firms pursue their technical and commercial interests). Its governance approach is also facilitative, involving a balanced blend of direction, participation, and voluntarism. For example, all of the strategic initiatives developed by Blade.org's principal office are intended to promote community development, while the costs and investments required to take advantage of those initiatives are largely borne by the member firms themselves. As the community-of-firms collaborative model spreads to other innovation-driven industries, we expect the facilitator role to continue to be important and therefore worthy of further research.

A final implication has to do with the anticipation of future organizational forms – identifying their purposes and characteristics as well as preparing for their arrival. The anticipation of new organizational forms can be approached by exploring the potential and limits of current organizational forms. As the newest form of organizing, the community-of-firms model needs to be examined in this regard. For example, a collaborative community of firms is a powerful means of producing a wide range of innovative products and services based on a single technology. Blade.org has already developed 60 unique solutions based on the blade processor technology, and the community's member firms are optimistic that many other solutions will follow. It is probably true, however, that the market and technical complementarities which are allowing Blade.org to grow will eventually reach their limit. At that point, Blade.org – or another community of firms that has evolved to the same stage – will begin to consider alternative organizational approaches that appear to be less restrictive in the development of new products and markets. Studying the community's adaptive response at that stage will be instructive in identifying the next organizational form that will emerge.

1.6 Conclusion

This chapter describes a new organizational form, a collaborative community of firms. The community-of-firms model is particularly well suited to the pursuit of continuous and complex innovation, and we expect to see this organizational approach used increasingly as firms come to realize the value to be gained from multi-firm collaboration both within and across industries. Blade.org, one of the pioneers of this new organizational approach, shows that a community of firms can be a successful means of innovation, and its organization design is worthy of study and imitation.

Acknowledgments We thank Lisa Galish of Blade.org for her guidance and support at several points in the research process and Stephan Ballinger, Dorthe Døjbak Håkonsson, and Raymond Miles for their helpful comments on an earlier version of the chapter. We are grateful for financial support provided by the Mellon Foundation and by Penn State's Smeal College of Business.

Appendix: Blade.Org Committees

1. Technology Committee – Focuses on the continued advancement of blade technology through forums and the establishment of common, industry-accepted semantics and taxonomy for blade system platforms.
2. Solutions Architecture Committee – Maintains a portfolio of blade solutions and is responsible for developing and maintaining the solution submission process, identifying new solution areas to create focus groups, soliciting solutions from member companies, and reviewing submitted solutions for approval.
3. Hosted Client Work Group – Develops business for Blade.org members through hosted client working group activities and the hostedclient.org web site, but because hosted client is not limited to blade technologies, its activities reach beyond blades and blade systems to encompass all aspects of hosted client solutions.
4. Power and Cooling Committee – Responsible for areas such as identifying power and cooling needs of the end user; mapping out elements that contribute to power and cooling efficiency and effectiveness from the processor to the power grid; and identifying power and cooling challenges and trends in order to propose new solutions.
5. Compliance and Interoperability Committee – Responsible for continually developing blade platform interoperability and accelerating customer adoption of industry-standard solutions.
6. Marketing Committee – Responsible for marketing and promotion of Blade.org activities including, but not limited to, the Blade.org web site, publicity programs and press releases, advertisements, developing consensus among member companies, and identifying strategic alliances for the organization.
7. SMB (Small and Medium Businesses) Committee – Responsible for increasing the adoption of blade-based technologies and solutions among small- and medium-size business firms by addressing their needs for simplicity and ease of deployment.
8. Membership Benefits Committee – Responsible for communicating to existing members the benefits of their respective membership level, providing members a feedback mechanism with regard to requested benefits, seeking new ways to increase member value, and working to attract new members to Blade.org.
9. Bylaws and Membership Committee – Ongoing mission to assist and advise the Blade.org Board of Directors on the implementation of bylaws and membership agreement amendments.

References

Aldrich HE (1979) Organizations and Environments. Prentice-Hall, Englewood Cliffs, NJ.
Aldrich HE, Ruef M (2006) Organizations Evolving. 2nd edn. SAGE Publishers, Thousand Oaks, CA.
Antorini YM (2005) The Making of a Lead User, Working Paper, Copenhagen Business School.
Bagozzi RP, Dholakia UM (2006) Open Source Software User Communities: A Study of Participation in Linux User Groups. Management Science 52: 1099–1115.
Baldwin C, Hienerth C, von Hippel E (2006) How User Innovations Become Commercial Products: A Theoretical Investigation and Case Study. Research Policy 35: 1291–1313.
Baumol WJ, Litan RE, Schramm CJ (2007) Good Capitalism, Bad Capitalism, and the Economics of Growth and Prosperity. Yale University Press, New Haven, CT.
Boden MA (1994) Dimensions of Creativity. The M.I.T. Press, Cambridge, MA.
Brown JS, Duguid P (1991) Organizational Learning and Communities-of-Practice: Toward a Unified View of Working, Learning, and Innovation. Organization Science 2: 40–57.
Brown JS, Duguid P (2000) The Social Life of Information. Harvard Business School Press, Boston, MA.
Brown JS, Duguid P (2001) Knowledge and Organization: A Social-Practice Perspective. Organization Science 12: 198–213.
Chesbrough HW (2006) Open Innovation: The New Imperative for Creating and Profiting from Technology. Harvard Business School Press, Boston, MA.
Chesbrough HW (2007) Why Companies Should Have Open Business Models. M.I.T Sloan Management Review 48: 22–28.
Clabby Analytics (2007). Blade.Org: The Snowball Effect, http://www.clabbyanalytics.com.
Cohen WM, Levinthal DA (1990) Absorptive Capacity: A New Perspective on Learning and Innovation. Administrative Science Quarterly 35: 128–152.
Dahl DW, Moreau P (2002) The Influence and Value of Analogical Thinking During New Product Ideation. Journal of Marketing Research 34: 47–60.
Demil B, Lecocq X (2006) Neither Market nor Hierarchy nor Network: The Emergence of Bazaar Governance. Organization Studies 27: 1447–1466.
Dyer JH, Singh H (1998) The Relational View: Cooperative Strategy and Sources of Interorganizational Competitive Advantage. Academy of Management Review 23: 660–679.
Ebadi YM, Utterback JM (1984) The Effects of Communication on Technological Innovation. Management Science 30: 572–585.
Franke N, Shah S (2003) How Communities Support Innovative Activities: An Exploration of Assistance and Sharing Among End-Users. Research Policy 32: 157–178.
Gassmann O (2006) Opening Up the Innovation Process: Towards an Agenda. R&D Management 36: 229–236.
Gulati R (2007) Managing Network Resources. Oxford University Press, New York.
Hargadon A (2003) How Breakthroughs Happen: The Surprising Truth about How Firms Innovate. Harvard Business School Press, Boston, MA.
Harhoff D, Henkel J, von Hippel E (2003) Profiting from Voluntary Information Spillovers: How Users Benefit by Freely Revealing Their Innovations. Research Policy 32: 1753–1769.
Harhoff D, Mayrhofer P (2007) User Communities and Hybrid Innovation Processes: Theoretical Foundations and Implications for Policy and Research, Working Paper, University of Munich.
Hult GTM, Hurley RF, Knight GA (2004) Innovativeness: Its Antecedents and Impact on Business Performance. Industrial Marketing Management 33: 429–438.
Hunt CS, Aldrich HE (1998) The Second Ecology: The Creation and Evolution of Organizational Communities as Exemplified by the Commercialization of the World Wide Web. In: Staw BM, Cummings LL (eds), Research in Organizational Behavior 20. JAI Press, Greenwich, CT, pp 267–302.
Huston L, Sakkab N (2006) Connect and Develop: Inside Procter & Gamble's New Model for Innovation. Harvard Business Review 84: 58–66.

Jeppesen LB, Frederiksen L (2006) Why do Users Contribute to Firm-Hosted User Communities? The Case of Computer-Controlled Musical Instruments. Organization Science 17: 45–63.
Jeppesen LB, Molin M (2003) Consumers as Co-Developers: Learning and Innovation outside the Firm. Technology Analysis & Strategic Management 15: 363–383.
Koerner BI (2006) Geeks, Wired, 14(2): 106–113.
Kogut B, Metiu A (2001) Open-Source Software Development and Distributed Innovation. Oxford Review of Economic Policy 17: 248–264.
Käser PAW, Miles RE (2002) Understanding Knowledge Activists' Successes and Failures. Long-Range Planning 35: 9–28.
Lakhani KR, Jeppesen LB, Lohse PA, Panetta JA (2007) The Value of Openness in Scientific Problem Solving, Working Paper 07-050, Harvard Business School.
Lave J, Wenger E (1991) Situated Learning: Legitimate Peripheral Participation. Cambridge University Press, Cambridge, UK.
Lee GK, Cole RE (2003) From a Firm-Based to a Community-Based Model of Knowledge Creation: The Case of the Linux Kernel Development. Organization Science 14: 633–649.
Lerner J, Tirole J (2002) Some Simple Economics of Open Source. Journal of Industrial Economics 46: 125–156.
Lilien GL, Morrison PD, Searls K, Sonnack M, von Hippel E (2002) Performance Assessment of the Lead User Idea-Generation Process for New Product Development. Management Science 48: 1042–1059.
Miles RE, Miles G, Snow CC (2005) Collaborative Entrepreneurship: How Communities of Networked Firms Use Continuous Innovation to Create Economic Wealth. Stanford University Press, Stanford, CA.
Miles RE, Miles G, Snow CC, Blomqvist K, Rocha HO, forthcoming, The I-Form Organization, California Management Review.
Palmer I, Dunford R (2002) Out with the Old and in with the New? The Relationship between Traditional and New Organizational Practices. International Journal of Organizational Analysis 10: 209–225.
Palmer I, Dunford R (2008) New Organizational Forms – the Career of a Concept. In: Barry D, Hansen H (eds), The SAGE Handbook of New Approaches in Management and Organization. SAGE Publishers, Thousand Oaks, CA, pp 567–569.
Powell WW (1996) Inter-Organizational Collaboration in the Biotechnology Industry. Journal of Institutional and Theoretical Economics 151: 197–215.
Powell WW, Koput KW, Smith-Doerr L (1996) Interorganizational Collaboration and the Locus of Innovation: Networks of Learning in Biotechnology. Administrative Science Quarterly 41: 116–145.
Pruegl R, Schreier M (2006) Learning from Leading-Edge Customers at The Sims: Opening Up the Innovation Process Using Toolkits. R&D Management 36: 237–250.
Raymond ES (1999) The Cathedral and the Bazaar. O'Reilly & Associates, Sebastopol, CA.
Ruef M (2002) Strong Ties, Weak Ties and Islands: Structural and Cultural Predictors of Organizational Innovation. Industrial and Corporate Change 11: 427–449.
Ryan RM, Deci EL (2000) Intrinsic and Extrinsic Motivations: Classic Definitions and New Directions. Contemporary Educational Psychology 25: 54–67.
Sawhney M, Prandelli E (2000) Communities of Creation: Managing Distributed Innovation in Turbulent Markets. California Management Review 42: 24–54.
Schumpeter JA (1934) The Theory of Economic Development: An Inquiry into Profits, Capital, Credit, Interest, and the Business Cycle. Harvard University Press, Cambridge, MA.
Shah S (2005) From Innovation to Firm Formation in the Windsurfing, Skateboarding, and Snowboarding Industries, Working Paper 05-010, University of Illinois.
Snow CC, Fjeldstad ØD, Lettl C, Miles RE (2009) Organizing Continuous Product Development and Commercialization: The Collaborative Community of Firms Model, Working Paper, The Pennsylvania State University.

Stabell CB, Fjeldstad ØD (1998) Configuring Value for Competitive Advantage: On Chains, Shops, and Networks. Strategic Management Journal 19: 413–437.
Thompson JD (1967) Organizations in Action. McGraw-Hill, New York.
von Hippel E (1988) The Sources of Innovation. Oxford University Press, Oxford, UK.
von Hippel E (2001) Innovation by User Communities: Learning from Open-Source Software. M.I.T. Sloan Management Review 41: 82–86.
von Hippel E (2005) Democratizing Innovation: Users Take Center Stage. The M.I.T. Press, Boston, MA.
von Hippel E (2007) Horizontal Innovation Networks – by and for Users. Industrial and Corporate Change 16: 293–315.
von Hippel E, von Krogh G (2003) Open Source Software and the Private-Collective Innovation Model: Issues for Organization Science. Organization Science 14: 209–223.
von Krogh G, Spaeth S, Lakhani KR (2003) Community, Joining, and Specialization in Open Source Software Innovation: A Case Study. Research Policy 32: 1217–1241.
von Krogh G, von Hippel E (2006) The Promise of Research on Open Source Software. Management Science 52: 975–983.
Ward TB (2004) Cognition, Creativity, and Entrepreneurship. Journal of Business Venturing 19: 173–188.
Wenger E (1998) Communities of Practice. Cambridge University Press, Cambridge, UK.
Zboralski K, Salomo S, Gemuenden HG (2006) Organizational Benefits of Communities of Practice: A Two-Stage Information Processing Model. Cybernetics and Systems 37: 533–552.
Zhongqi J, Hewitt-Dundas N, Thompson NJ (2004) Innovativeness and Performance: Evidence from Manufacturing Sectors. Journal of Strategic Marketing 12: 255–266.

Chapter 2
Network-Level Task and the Design of Whole Networks: Is There a Relationship?

Patrick Kenis, Keith G. Provan, and Peter M. Kruyen

Abstract This chapter explores a key argument of structural contingency theory in the context of whole networks of organizations. Specifically, we examine the relationship between the task such networks perform and their design, which we operationalize as one of three forms of governance. Based on an extensive review of the literature on whole networks, we conclude that there is no clear relationship between network-level task and network design. We offer a number of explanations that might account for this finding. Our conclusions help to advance theory and practice on the design of whole networks of organizations.

Keywords Whole networks · Network design · Network governance · Network tasks

2.1 Networks as Production Systems

Recently, the idea of networks as production systems has gained attention in the literature on organizations (cf. Miles et al. 2005; Berkhout and De Ridder 2008). The focus of this perspective is on the level of the whole network, arguing that organizations collaborate in networks to produce joint or collaborative outcomes. These are different outcomes than those that individual actors in networks pursue. Network-level tasks include integration of critical services to vulnerable populations like people with AIDS or those with serious mental illness, regional economic development, developing or producing an expensive and highly complex project like building a bridge or making a major movie, or responding to natural or man-made disasters. Advocates of networks as production systems argue that networks are an innovative way of getting complex jobs done, especially those that are beyond the capacities of a single organization.

P. Kenis (✉)
TiasNimbas Business School, Tilburg University, Tilburg, The Netherlands
e-mail: p.kenis@uvt.nl

A. Bøllingtoft et al. (eds.), *New Approaches to Organization Design*, Information and Organization Design Series 8, DOI 10.1007/978-1-4419-0627-4_2,

The goal of this chapter is to contribute to the study of the design of such whole networks. From descriptive case studies (cf. Agranoff and McGuire 2003; Goldsmith and Eggers 2004; Huxham and Vangen 2005; Van Bueren et al. 2003), we learn that whole networks come in different forms. The question to be explored here is, why is this the case? What explains the form of whole networks?

From case studies we know that whole networks are a prevalent phenomenon, but conceptual, analytical or explanatory studies of whole networks are scarce. In a recent overview of the literature, Provan et al. (2007) identified only 26 analytical studies in which the whole network was the unit of analysis. Various reasons can be identified why analytical studies of whole networks are rare. One of the reasons may simply be that organizational scholars are used to and trained to study organizations rather than multi-organizational arrangements (Salancik 1995). In addition, developing a deep understanding of multi-organizational networks requires costly and extensive data collection. Moreover, the study of whole networks seems often to be confined to fields in which the obvious outcomes are collective or public goods (i.e., public policies, community health, emergency response to disaster, and the like). For instance, Provan et al. (2007) found that of the 26 whole network studies identified, 14 of them were in the health and human services sector. The assumption is that public and nonprofit organizations are supposed to produce collaborative or collective goods whereas private (for-profit) organizations have as their sole objective the maximization of individualistic benefits. The idea that customer value can also, and often only, be created through a whole network is still an underrated and under-explored idea (for exceptions see Prahalad and Ramaswamy 2004; Miles et al. 2005).

We start our thinking about whole network design from a structural contingency approach, given the dominance of this perspective at the organizational level of analysis. From this perspective a relationship between the tasks to be performed by the system and the work design of the system is expected.

Similar to what Woodward (1965), Burns and Stalker (1966) and Thompson (1967) did at the organizational level, we first introduce a typology of whole network governance. The logic is that one specific form of network design, operationalized here as the form of network governance, is better than another for accomplishing a particular type of task. Second, we describe our research methodology for examining this basic contingency argument. Third, after developing a typology of network-level tasks based on what we found in the empirical literature, we analyze whether there is a relationship between the main forms of network governance and network-level tasks. Finally, we discuss the consequences of our findings.

2.2 Forms of Network Governance

Our discussion of network governance design forms is based on two basic premises. The first is that whole networks are a form of governance different from either a market or a hierarchy (see Williamson 1973; Powell 1990; Castells 2000; Provan and Kenis 2008). Second, network governance, as is the case with markets and hierarchies, can take wide variety of design forms. Our focus in this chapter is on the second point. Network governance is not a single approach but appears in different forms.

Whole networks, just like organizations, need to be structured in order to succeed in fulfilling their network-level task (Klijn 2005; Provan and Kenis 2008). For example, structures need to be in place to encourage network members to engage in collective and mutually supportive action, to address conflicts, and to ensure that resources are acquired and utilized efficiently and effectively. We draw on recent work by two of the authors of this chapter (Provan and Kenis 2008), who propose three basic network governance models: participant or shared governance, lead-organization governed, and network administrative organization governed. Each of these models differs in their basic structure.

2.2.1 Shared Governance Form

The simplest form of a network is one that has shared participant governance. Networks having a shared governance form consist of multiple organizations that work collectively as a network but with no distinct governance entity. Governance of collective activities resides completely with the network members themselves. In this model it is the network participants themselves that make all the decisions and manage network activities. There is no distinct, formal administrative entity, although when there are more than a handful of network participants, some administrative and coordination activities may be performed by a subset of the full network.

2.2.2 Lead Organization Form

In the lead organization model, network members all share at least some common purpose (as well as maintaining individual goals) and they may interact and work with one another. However, all activities and key decisions are coordinated through and by one of the members, acting as a lead organization. This organization provides administration for the network and/or facilitates the activities of member organizations in their efforts to achieve network goals.

2.2.3 Network Administrative Organization Form

The basic idea of the network administrative organization (NAO) governance model is that a separate administrative entity is set up specifically to manage and coordinate the network and its activities. Like the lead organization model, the NAO plays a key role in coordinating and sustaining the network. Unlike the lead organization model, however, the NAO is not another network member, providing its own set of services or production tasks. Instead, the NAO is established with the exclusive purpose of network governance. It may be a government entity, or more likely, a nonprofit, which is often the case even when the network members are for-profit firms (cf. Human and Provan 2000).

NAO's may be relatively informal structures, consisting of a single individual who acts as the network facilitator or broker, or they may be much more formalized and complex, consisting of an executive director, staff, and board (consisting of network members) operating out of a physically distinct office.

2.3 Research Methodology

To build a systematic overview of network tasks and to study how the type of task is related to the network design/governance mode, a *systematic literature review* was conducted. According to Mulrow (1994), this is the best research method for identifying and categorizing the extant literature on a certain topic. It differs "...from traditional narrative reviews by adopting a replicable, scientific and transparent process, in other words a detailed technology, that aims to minimize bias through exhaustive literature searches..." (Tranfield et al. 2003: 209). Such a review includes the systematic collection, analysis and synthesis of data available in the extant literature.

We aimed at identifying as many studies as possible, and therefore, searched 10 established academic databases varying from ABI/Inform to Wiley InterScience. Because descriptions of network tasks and designs are not presented in a systematic way, we started by including all the literature that in some way referred to inter-organizational networks, even if no explicit reference was made to network task or network design. Search terms included: whole, (inter)organizational, (inter- and multi-) firm, public, policy, and business networks, consortia, clusters, alliances, partnerships, collaborative governance, and service integration. Moreover, the search in all databases was restricted to peer-reviewed journals to guarantee a minimal quality (Finfgeld 2003). We made an initial assessment on the basis of the abstract and continued only with those papers that had an explicit focus on inter-organizational networks. This first step resulted in a list of 436 articles.

In a second round, we read all the remaining papers thoroughly and kept only papers in which the central object of study was unambiguously on inter-organizational networks. Moreover, only those papers were included in which empirical evidence was provided to support the claims made by the researchers (Finfgeld 2003). We kept only those papers in our sample in which (1) a description was given of the task(s) conducted by the network(s) studied; (2) a description was given of the network design(s); and (3) a statement was made about the success or failure of the network in executing its task(s). This second round resulted in a final set of 55 papers, which forms the basis of the remainder of this chapter. (A list of the articles reviewed is available upon request).

2.4 Network-Level Tasks

In analyzing the literature we brought dispersed findings together, interpreted them and transformed them into a new whole (see for a description of this method Sandelowski et al. 1997; Finfgeld 2003). Thus, inferences are based on our own interpretation of the data rather than on what the authors said. From this exercise, two analytical dimensions emerged along which network tasks could be described. The first dimension differentiates between exploitation and exploration. Exploration can be defined as searching for new ways to tackle existing or new problems and challenges – the development of new products, services, policies, etc. Exploitation is the actual delivery or production of goods, services, policies, etc. This distinction is commonly used at the organizational level of analysis regarding organizational learning (March 1991) but it is also relevant at the network level. A second dimension that emerged from our research of the literature distinguishes between unambiguous and ambiguous tasks. In unambiguous tasks member organizations know with a high degree of certainty what is expected from them in order to accomplish the collective network task. In contrast, when tasks are ambiguous, member organizations are far more uncertain about how their actions might contribute to the collective task. Dichotomizing and crossing both dimensions lead to four broad categories of tasks (see Fig. 2.1).

Cell I. This category contains networks in which the production process is standardized and member organizations know what is expected as their contribution to the network task and how their part of the collective task should be accomplished. Examples include the production of standardized products such as timber (Gellert 2007), frozen meat (Hunter 2005), and automobiles (Dyler 1996). Also, the production of standardized services such as school curricula to educate nurses (Haas et al. 2002) and the provision of public network services providing electricity (Hughes 2005). Finally, the provision of standardized network services for network members fits cell I. Examples include sharing production capacity and skills in a surgical

	Unambiguous task	**Ambiguous task**
Exploitation task	**I** (e.g. auto manufacturing, electricity)	**II** (e.g. mental health care)
Exploration task	**III** (e.g. developing common standards)	**IV** (e.g. development of industry clusters; high tech innovation)

Fig. 2.1 Classification of network-level tasks

instruments cluster (Navdi 1999), disposal of polluted water in the production of leather (Kennedy 1999), or collecting, comparing, and sharing uniform data on infectious diseases (Parkinson et al. 2008).

Cell II. This category includes the delivery of network products in a predictable non-innovative way, but where outcomes are vague and difficult to measure and where the production process itself is unstandardized or has to be adjusted or adapted on a case-by-case basis. This includes the provision of a combination of social, mental, and physical health services to specific groups such as mental health clients (Provan and Milward 1995), adult, substance-abusing female offenders (Townsend 2004), or homeless people (Gordon et al. 2007). In the provision of these services, standard methods of treatment are usually applied but the specific needs of clients often vary and it is unclear whether and when the service is delivered successfully, can be terminated, or whether it has to be continued. As a consequence, it is unclear what specific contributions of member organizations are required to accomplish the collective task successfully.

Cell III. This category includes various types of tasks which have the ultimate purpose of developing one common outcome or standard. Examples include technical innovations, such as the development of a DVD-standard (De Laat 1999); the improvement of a whole national production chain (timber in South Africa: Bessant et al. 2003); or policy developments such as the regulation of specific financial products (Faerman et al. 2001). This category also includes (mutual) learning tasks which aim at finding a best practice solution with respect to core activities such as quality management systems in the production of mass-consumer goods (Altenburg and Meyer-Stamer 1999) or medical services to pregnant and nurturing mothers (Valadez et al. 2005).

Cell IV. This category includes network tasks that focus on the continuous development (Miles et al. 2005) of a broad field (e.g., an industry, policy area, or service delivery system). Examples can be found in the development of geographic areas or industry clusters, for example, the toy industry in Spain (Holström 2006), and the high-tech industry in Finland (Jauhiainen 2006). Also in this category belongs the development of a broad market-reform agenda (in Mexico, Salas-Porras 2005; and Western-Africa, Brinkerhoff 1999). Other examples include learning about strategies to fight multinational corporations by the anti-sweatshop movement (Connor 2004) and the provision of human and social services to HIV/AIDS patients (Altenstetter 1994). In these examples, the solution to the network task is not only difficult to define and measure, but the means to accomplish network goals are not clear and may not even exist. It is up to network members to work together to find both the methods to address complex problems and even to define the problem itself.

In conclusion, the literature review demonstrates that networks conduct different types of network tasks that can be broadly categorized in one of four ways based on the exploitation/exploration dimension and the degree of ambiguity. The question now arises whether effective task accomplishment is related to specific network designs, consistent with structural contingency argument developed for the organizational level (see Donaldson 1996).

2.5 Network-Level Task and Network Design

In what follows we explore whether working on a specific type of network-level task is related to a specific network design, as operationalized by one of the three distinct network governance forms described earlier.

Cell I. We found that the tasks in this cell are performed by networks having all three design forms. There were several instances of a lead organization. This was the case in the manufacturing of automobiles (Dyler 1996), aerospace products (Moffat and Archer 2004), and the provision of nursing education in a standard curriculum (Haas et al. 2002). In all these networks there was one member organization which conducted a large amount of the work flow. The lead organization assembled the products or provided a large amount of the schooling, while the other members delivered sub-parts (e.g., computer chips by manufactures or internships by hospitals). As Alter and Hage (1993) proposed, because this organization conducts a large part of the task, it dominates the network. The lead organization is able to measure or monitor whether the task of the network is fulfilled and, in turn, can give specific directions to other participating organizations when, for example, the quantity or quality of their output has to be adapted. As a consequence, member organizations know exactly what is expected of them. Moreover, the lead organization keeps an eye on other network members' output and may intervene when they underperform. As a result, the performance of the whole network output is strengthened, which in turn leads to the increased benefits for all participants.

The empirical literature also revealed, however, that in those cases where a large number of organizations are involved and where not one single organization conducted a central part of the work, a separate entity tends to be constructed in order to coordinate the network. Such a NAO governed network was found in the provision of nurse education through the collaboration of various schools and hospitals (Allen et al. 2007). A NAO was also present in the "canonical district" of Prato in which the *impannatore* governs the whole production process, from the purchase of raw materials from various producers, with which long-standing relationships were held, to the marketing of the final products. The NAO, however, owned no physical assets itself (Paniccia 1998). Moreover, it was also used in the production of frozen meat in New-Zealand (Hunter 2005) and by the Indonesian government in the case of timber production (Gellert 2007). Lastly, we found a NAO in the "production" of fair college sport competition in the USA (Stern 1979).

The reason for using a NAO in the production of these goods is probably because without such an entity the network would not exist in the first place. No organization conducts a central part of the work flow and every member is probably too small to have enough resources to coordinate the network as a whole. Besides, it is unlikely that other member organizations would allow their production to be coordinated by another network organization, which could be regarded as a potential competitor. Moreover, a shared governed network would not be effective in either of these cases. This is mainly due to the size of the network. It would be too time consuming to meet periodically and govern the network collaboration. In contrast, a

NAO can, as an independent entity, monitor the quantity and quality of the various network-members and prevent members from undermining the collective task and resolve conflicts. Examples of this might include supplying more wasted material than members had agreed upon, in the case of the common waste disposal factory (Kennedy 1999), or having disputes about the rules of the game in the college sport competition network (Stern 1979). In the case of the trade association in the surgical instrument cluster in Pakistan (Navdi 1999), the authors concluded that the network was unable to enforce quality standards, limit price competition, and apply sanctions against misbehaving members because it was self-governed by network participants and had no separate governance authority.

We did, however, find a number of shared governance networks in this task category. In all these cases the networks were small. For example, in the network in which three organizations collaborated to provide electricity (Hughes 2005), each member sent several representatives to periodic committee meetings where members negotiated the selling and buying prices that each would charge when supplying energy to the network. It was also the case in the collaborative buying of energy by 15 manufacturing firms in England (Hanna and Walsh 2004). It is probably because of the small size of these networks that monitoring each others' contribution could be based on multilateral observation and trust.

Based on our review, we conclude that when a network has an unambiguous exploitation task, the network can have any of the three design forms and that design is more likely to be affected by work flow and trust conditions.

Cell II. As with cell I, we found examples of all three design/governance forms. In four rural health care delivery networks in Nebraska, a separate organization (i.e., a NAO) was set up with official coordinators and a board, consisting of hospital administrators and which made strategic plans and managed the budget (Schumaker 2002). The same holds for health delivery networks in New Hampshire in which a variety of members were involved including hospitals, health and social service organizations, schools, police and fire departments, businesses, and municipalities. In these networks a designated coordinator oversaw day-to-day operations of site activities and kept in contact with coordinators of other health delivery networks (Kassler and Goldsberry 2005).

But examples of successful lead governed networks are also abundant in this category. For example, Townsend (2004) describes four networks that provided substance abuse treatment to adult, substance-abusing female offenders. Two of these networks were led by a court and two others by an adult probation department. Furthermore, Atkinson and Gonet (2007) described a network of five organizations that provided statewide post adoption services. This network was managed by a private, nonprofit organization which provided a wide range of adoption services across the state.

Theoretically speaking, it is reasonable to expect a NAO or lead organization in this task category. Because the network task is ambiguous, member roles, activities, and contributions are often not clear. The NAO or lead organization can develop a general mission and standardized procedures under which resources are shared and exchanged and is at least able to monitor whether members follow the specified procedures. Moreover, this organization, can "sell" the network's mission to the general world to acquire new collective resources.

The reviewed empirical literature does not, however, give a clear idea whether or under what conditions the NAO or the lead governed network is most effective. Both seem to have specific advantages. The lead organization is capable of providing a considerable part of the service it knows as well as additional services or products needed for a well functioning network. The NAO, on the other hand, has as an advantage in that it is an independent entity, allowing member organizations to feel that they are not dominated by a peer organization. Based on our review, the choice for a lead or NAO governed network seems to be made instead on historical grounds (i.e., which organization provided a large portion of the service before the network was established) and/or mandates from the funding agencies. On the basis of the article by Provan and Kenis (2008) we could argue that in networks in which trust is traditionally high, a lead organization is more likely and that in organizations with low or moderate trust between the network partners, a NAO is more likely.

But we also found that a number of successful shared, participant governed networks in this category. In the case described by Foster-Fishman et al. (2001), the human service network was managed by 32 organizations and the various service delivery teams had nearly twenty members. In another network described by Norman and Axelsson (2007), five organizations collaborated in a shared governance network to provide rehabilitation services. Finally, in a shared governance HIV/AIDS service-delivery network (Takahashi and Smutny 2002: 177–178), managers of the three member organizations frequently met and "discussed fundraising strategies, overall partnership goals, crises, conflict and miscommunication, and longer range plans for expansion of partnership membership." According to the authors of these various studies the networks succeeded because of the development of high goal consensus and a feeling of trust among members. For example, the service network described by Foster-Fishman et al. (2001) developed procedures to deal with conflicts, while the three member organizations of the HIV/AIDS network (Takahashi and Smutny 2002) first engaged in a "honeymoon period" to get to know each other and build trust before the network was formally established. The rehabilitation project in Sweden was assisted by a third-party facilitator, joint seminars were held, and informal contacts were established to develop a common language and mutual-understanding.

Cell III. Tasks in this cell tend to be performed most often through a participant governed structure, but our review also found cases where a lead organization and NAO design was used. For instance, a shared, participant governed network was applied by the schools that formed networks in England to develop the curriculum, standardized appraisal procedures, preparation for inspection, and calendar coordination. They met regularly and these meetings were chaired on a rotating basis, usually by each school in turn (Busher and Hodgkinson 1995). These schools did not perceive each other as competitors, as was the original intention of the deregulation policy of the government, but rather, saw cooperation as a better method to develop and implement improvements in the school system.

Although most cell III tasks were performed by shared-governance networks, we also found a number of lead organization networks. For example, the network described by McKenna et al. (2003: 384) of seven public safety and health organizations to develop "a reliable, easy to use electronic surveillance system for bioterrorism and other infectious mass disease emergencies" was led by the local health authority responsible for all public health activities, including communication and control. Also, Moffat and Archer (2004) described a network in which innovations in the production of mature electric goods were carried out successfully under the coordination of a design house. The same holds for the two networks presented by Miles et al. (2005) in which BMW and General Electric each collaborated with main suppliers and customers to improve car seats (BMW) or the whole production chain (General Electric). A final successful lead governed network in this category is the network in which eight organizations, two government agencies and six large financial institutes jointly developed regulations for an innovative financial product (Faerman et al. 2001). Although being governed by one of the involved government agencies, it was facilitated by an independent consultant, an attorney of a law firm. Because of his independence, his knowledge of the sector and institutions involved, and the participating institutions trusting him in keeping company information secret, he was able to keep the organizations together despite the existing internal disagreements. In contrast, both the development of the Bluetooth technology (Rice and Juniper 2003) and the initial attempt to regulate package waste disposal (Nunan 1999) failed, according to the authors, because they were coordinated (and dominated) by a small number of organizations.

Finally, we found several instances in the literature in which an unambiguous development/exploration task was governed by a NAO. For instance, Von Malmborg (2007) studied a network in which 16 small and medium sized enterprises collaborated to find sustainable solutions to improve the production process. Four government agencies and 12 consultants were also involved and the entire network was coordinated by a member of one of the government agencies. This person organized thematic meetings each month, provided funding, and functioned as a communication channel without having any specific knowledge about the production processes of member firms. The network examined by Bessant et al. (2003) was also governed by a NAO. This network's purpose was to develop the whole timber production value chain in South-Africa by bringing together a number of timber growers, sawmills, timber production manufacturers, government agencies, the export council, and research institutes. Although sub-committees were coordinated by a person from a firm in the value chain, the network as a whole was coordinated by action researchers from a university.

Occasionally, such NAOs seem to fail, however. For example, the weekly meetings of a network in England directed at training new entrepreneurs, having approximately 50 organizations as members, was led by employees of a Training and Enterprise Counsel (Huggins 2000). This network was judged to be a failure by member organizations because of, among other things, a lack of professionalism by these independent facilitators.

The arguments for having a shared, participant governed structure, lead organization, or NAO form run more or less parallel to the reasons given in the previously discussed cells. Whenever one organization conducts a large part of the development task, it knows which contributions are needed from other organizations and it is able to monitor the quality of the contributions. When there is not such an organization conducting a central part of the development process, however, it is more likely that a NAO or a shared governance network design form will be used. The choice between these two forms is probably most related to whether or not the participants see each other as competitors. When they do *not* regard themselves as competitors it is more likely that they will organize the network themselves and trust each other to govern collaboratively (i.e., shared, participant governed). When, on the other hand, they perceive each other as competitors, a NAO network becomes more likely.

Cell IV. Finally, in the cases where network tasks require exploration and are ambiguous, all three network design forms occurred. An example of a self-governed network within this category is the network in Mexico described by Salas-Porras (2005) in which a large number of business associations, charitable, religious, educational and right-wing political institutions advanced a market-reform agenda. Here, several sub-networks were formed in which member organizations with similar interests worked together closely to explore and exploit specific parts of the agenda. The full network of participating organizations did not meet collectively to discuss different matters or to develop a shared agenda.

Another example of a shared governance design is the international network of organizations which forms the anti-sweatshop movement (Connor 2004). The organizations involved in this network participated relatively equally, engaging in multiple, often temporary sub-networks. Collectively, they were able to facilitate innovation and adaptive learning for the network as a whole.

A lead-governed network was present in the Canadian network which opposed the Canada-US and North American Free Trade Agreements (Huyer 2004). This network was made up of labor unions and civil society groups (e.g., churches, women's movement, environmental movement, cultural and social justice groups). Although the network was based on equal participation, it was managed by "representatives or leaders of the key member organizations of which labor played a high-profile role" (Huyer 2004: 53). Although the network did not directly succeed, it nevertheless forced the ruling conservatives to call an election before NAFTA was implemented and members still collaborate in anti-free trade activities (e.g., joint demonstrations, provision of information). A lead organization governance design was also present in a local network of business organizations and government agencies to improve social and community issues, like cleanness, community safety, and policing (Rogers and Anderson 2007). Each month these firms met and were facilitated by a project manager who was an employee of the government agency responsible for community development.

Finally, a number of examples can be found in the economic cluster literature in which member organizations only interact with each other through a hub-organization, which acts as a NAO. This hub-organization provides technical expertise and administrative support to individual participants, and engages in co-innovation with a limited number of member organizations. What makes this type of network structure a joint production system is that the hub-organization disseminates valuable information acquired by some network members to other network members, which can be used to stimulate innovation, thereby strengthening the competitiveness of the entire cluster. For instance, in the Canadian automotive parts industry, the union, besides having its own research facilities, also kept manufacturers informed about what was going on in other firms and which innovations had been implemented (Rutherford and Holmes 2007). The same hub-function was fulfilled by the Institute for Toys in the Spanish town of Ibi, which, among other things, "offers market information, advice on product development and manufacturing" (Holmström 2006: 495) and represents the industry at trade fairs. In the Spanish ceramic tile district, there are a number of hub-firms (local university, trade associations) that connect the cluster with other industries, countries, and knowledge forums, while at the same also engaging in R&D activities with individual businesses (Molina-Morales 2005).

In conclusion, our review indicates that in this task category, as with the other cells of our model, no clear relationship could be found between network task and network design, operationalized as network governance form. Instead, it appears that trust plays a prominent role. In general, if trust is high between network members, they are able to function in shifting coalitions in order to perform the collective network task without the presence of a specific entity responsible for governing the network. However, when trust is only modest, which was the case in the cluster examples, a NAO becomes functional in order to govern the network. And when there is one organization (or a small number of organizations, as was the case in the NAFTA example) which already performs a large amount of the network task, this organization is likely to take the lead.

2.6 Discussion

This chapter has been an attempt to examine, based on a review of the literature on whole networks, the relationship between the tasks performed by networks and their design. An important conclusion appears to be that networks are used for a wide range of tasks. This contrasts somewhat with most of the network literature which often claims that networks are a new form of governance designed to accomplish a specific type of task, typically requiring the rapid transfer of knowledge and information across organizational boundaries (cf. Castels 2000). This is the type of task that fits Cell IV in our typology but it is clear from our literature review that a wide range of task activities are being accomplished through whole networks, reflected in the number of studies categorized in Cells I, II, and III.

Probably the most striking conclusion of our literature review is that there does not appear to be a relationship between type of task and network design. This is in contrast to what one would expect on the basis of structural contingency theory. In what follows, we will critically reflect on this finding and suggest possible explanations.

First, it could be that variables other than type of task are more important for explaining network design. The literature we reviewed provides some evidence for this assumption. While not the focus of our research, two critical factors regularly appeared to have some explanatory power; namely, the level of trust between network member organizations and whether the central work flow was already controlled and managed prior to network formation by and through one or a small group of organizations. The importance of these factors might be related to a fundamental aspect of organizational networks, which is the fact that they are production forms that consist of independent organizations. As a consequence, trust and the resolving of competitive tensions among network members may be more critical for success in accomplishing network-level tasks than the task itself.

Another variable that might help explain network design is the size of the network. Size has been widely discussed by contingency theorists as an important factor for explaining organization structure (cf. Kimberly 1976), but not for explaining network structure. We observed that the fewer organizations involved, the more likely it was that the network was self-governed. Shared governance leaves control over the functioning of the network to the network members themselves. In a small network it would be highly inefficient to have all communication and network-level decisions go through a lead organization or NAO. For example, for conflict resolution, face-to-face communication might be functional in a small network, while it could be a bureaucratic burden in a lead organization or NAO. In contrast, in larger networks, shared governance might lead to a situation where members start to ignore network management issues or spend too much time trying to coordinate across a large number of organizations. In such a case, a NAO renders collective decision making unnecessary. Members only have to interact with the NAO to coordinate network-level activities. These arguments are consistent with what Provan and Kenis (2008) have already proposed in their article on network governance.

Second, it is possible that network design parameters other than governance mode are related to the task of the network. These could be, for example, the frequency of interactions among individual members or the density of interactions across the whole network. It may be that the more ambiguous the task, the more network members must interact to accomplish the task successfully. However, the literature reviewed did not reveal sufficient information on the importance of such other design parameters.

Third, and in line with our previous point, we were completely dependent on the quality and extensiveness of reporting by the authors of the literature we reviewed. As a result, coding sometimes turned out to be quite difficult. Often it was not easy to understand exactly what task the network had to accomplish and whether or not the task was actually being accomplished successfully. In addition, since governance was not the focus of most of the studies surveyed, the exact form of governance being used was sometimes difficult determine. All these problems resulted in coding that was not always as objective as we would have liked. However, in the absence of a single large scale comparative study of whole networks utilizing a single and consistent method for data collection, the literature review approach we used is all that is available.

2.7 Contributions

Finally, we would like to formulate how our research can contribute to future thinking about organizational design. First, there is the question of the prevalence of whole organizational networks as a way to create value. More empirical research is needed to demonstrate how dominant this form of organization is and whether and why we can expect an increase in the prevalence of this governance form. We also need to know more about whether whole networks are really different from those organizational forms that previously have been described as "loosely coupled systems" (Weick 1976), political systems, network organizations, adhocracies, and the like. We think they are different because networks are goal-directed multi-organizational production systems organized to accomplish a task, rather than loose coalitions that are formed serendipitously. In addition, whole networks are composed of independent, sovereign organizations that can voluntarily leave the network. These systems require specific mechanisms for governance, which we have used here as our operationalization of design. But more research is needed to make explicit what these differences entail in terms of the functioning of these multi-organizational forms.

Assuming that whole networks are indeed becoming more prevalent and that they are a different, non-hierarchical form of organizing, the appropriateness of existing organizational theories for explaining networks becomes an important issue. For the most part, the organizational design literature has focused on individual organizations, while theories of networks and network structure have mainly focused on dyadic and ego-centric network ties, seldom focusing on whole networks as the unit of analysis. Thus, a new way of theorizing about networks based on the design of the whole system as a mechanism for accomplishing a collective task is needed. Our work has provided evidence that conventional theories, especially arguments based on structural contingency theory, do not have predictive value for whole networks. Our conclusions provide only limited evidence, however, concerning how network design might best be explained. Future research is needed to identify, through new empirical research or through systematic literature reviews like this one, those factors that do have predictive power for whole network design. Our findings provide suggestive evidence that network size (in terms of the number of participating organizations) and trust among the participating organizations seem to explain, at least to some extent, the organizational design of whole networks, consistent with recent conceptual work by Provan and Kenis (2008).

Another way of thinking about this issue could be to study whether the emergence and importance of whole networks and other "new" forms of organizing means that classical organization design approaches are relevant any more. In particular, it may be that traditional design approaches and theories are "backward oriented," focusing on what has worked in the past. The emergence of whole networks as a way of doing work could be an indication that having a more "forward orientation" is considered important by more and more organizations. If this is the case, then researchers might analyze if the orientation toward time (backward versus forward orientation) is predictive for the design of whole networks. This point appears to be consistent with what Huber (see Chapter 1) refers to as "designs looking ahead." It may be that the design of whole networks is determined by the extent to which participating organizations need to be part of a system that has the capacity to anticipate future needs, rather than reacting to current task demands. One clear advantage of whole networks, as compared with more hierarchical forms, is that they are able to "shoot at a moving target." Most organizations have a top-down structure with limited points of access for critical information, and thus, they have limited ability to recognize the need for change. In contrast, networks have many points of access to different sources of different information, which can be readily disseminated among members, allowing them not only to anticipate the need for change but even to shape the change itself (Burt 2005).

From a practice perspective, our findings demonstrate that network organizers and managers may not need to design network governance structures based on characteristics of the task the network is trying to accomplish. Networks must be governed if they are to be effective in achieving network-level goals. However, from a review of the literature on the topic, it appears that task characteristics may be far less critical than issues like network size, the trust level among participants, or the need to be responsive to change. Network managers and organizers must recognize these factors and respond accordingly.

These alternative conclusions are, of course, somewhat speculative, since they are grounded in the logic of networks discussed in the literature, as opposed to being based on our findings. It is obvious that more research is needed to test these assumptions directly. Despite this shortcoming, we hope that the analysis presented in this chapter provides some clear direction for future research as well as some motivation to initiate further study on the topic. We believe that there is a bright future for the study of network design, which has been a relatively neglected aspect of the important field of organization design.

References

Agranoff R, McGuire M (2003) Collaborative public management: New strategies for local governments. Washington, DC: Georgetown University Press.

Allen P, Schumann R, Collins C, Selz N (2007) Reinventing practice and education partnerships for capacity expansion. Journal of Nursing Education 46: 170–175.

Altenburg T, Meyer-Stamer J (1999) How to promote clusters: Policy experiences from Latin America. World Development 27: 1693–1713.

Altenstetter C (1994) European Union responses to AIDS/HIV and policy networks in the pre-Maastricht era. Journal of European Public Policy 1: 413–440.

Alter C, Hage J (1993) Organizations working together. Newbury Park, CA: Sage Publications.
Atkinson A, Gonet P (2007) Strengthening adoption practice, listening to adoptive families. Child Welfare 86: 87–104.
Berkhout G, De Ridder W (2008) Vooruitzien is regeren: Leiderschap in innovatie. [Foresight is the essence of government: Leadership in innovation]. Amsterdam, the Netherlands: Pearson Education Benelux.
Bessant J, Kaplinsky R, Morris M (2003) Developing capability through learning networks. International Journal of Technology Management and Sustainable Development 2: 19–38.
Brinkerhoff DW (1999) State-civil society networks for policy implementation in developing countries. Policy Studies Review 16: 123–147.
Bueren E van, Klijn EH, Koppenjan JFM (2003) Dealing with wicked problems in networks: Analysing an environmental debate from a network perspective. Journal of Public Administration Research and Theory 13: 193–212.
Burns T, Stalker GM (1966) The management of innovation. London, Great Britain: Tavistock Publications.
Burt RS (2005) Brokerage and closure: An introduction to social capital. Oxford, NY: Oxford University Press.
Busher H, Hodgkinson K (1995) Managing interschool networks: Across the primary/secondary divide. School Organisation 15: 329–340.
Castells M (2000) The information age: Economy, society and culture, vol. 1. Malden, MA: Blackwell Publishers.
Connor T (2004) Time to scale up cooperation? Trade unions, NGOs, and the international anti-sweatshop movement. Development in Practice 14: 61–70.
de Laat PB (1999) Systemic innovation and the virtues of going virtual: The case of the digital video disc. Technology Analysis & Strategic Management 11: 159–180.
Donaldson L. (1996) The normal science of structural contingency theory. In: Clegg SR, Hardy C and Nord W (eds), The handbook of organization studies. London, Great-Britain: Sage, pp 57–76.
Dyler JH (1996) Does governance matter? Keiretsu alliances and asset specificity as sources of Japanese competitive advantage. Organization Science 7: 649–666.
Faerman SR, McCaffrey DP, Slyke DM van (2001) Understanding interorganizational cooperation: Public-private collaboration in regulating financial market innovation. Organization Science 12: 372–388.
Finfgeld DI (2003) Metasynthesis: the state of the art – so far. Qualitative Health Research 13: 893–904.
Foster-Fishman PG, Salem DA, Allen NA, Fahrbach K (2001) Facilitating interorganizational collaboration: The contributions of interorganizational alliances. American Journal of Community Psychology 29: 875–905.
Gellert PK (2007) From managed to free(r) markets: Transnational and regional governance of Asian timber. The Annals of The American Academy of Political and Social Science 610: 246–259.
Goldsmith SE, Eggers WD (2004) Governing by network. Washington, DC: Brookings.
Gordon AJ, Montlack ML, Freyder P, Johnson D, Bui T, Williams J (2007) The Allegheny initiative for mental health integration for the homeless: Integrating heterogeneous health services for homeless persons. American Journal of Public Health 97: 401–405.
Haas BK, Deardorff KU, Klotz L, Baker B, Coleman J, DeWitt A (2002) Creating a collaborative partnership between academia and service. Journal of Nursing Education 41: 518–523.
Hanna V, Walsh K (2004) How to co-operate for competitive advantage. Engineering Management Journal 14: 28–31.
Holmström M (2006) Globalisation and good work: Impiva, a Spanish project to regenerate industrial districts. Tijdschrift voor Economische en Sociale Geografie 97: 491–502.
Huggins R (2000) The success and failure of policy-implanted inter-firm network initiatives: Motivations, processes and structure. Entrepreneurship & Regional Development 12: 111–135.

Hughes TP (2005) From firm to networked systems. Business Historical Review 79: 587–593.
Human SE, Provan KG (2000) Legitimacy building in the evolution of small-firm networks: A comparative study of success and demise. Administrative Science Quarterly 45: 327–365.
Hunter I (2005) Commodity chains and networks in emerging markets: New-Zealand, 1880–1910. Business Historical Review 79: 275–304.
Huxham C, Vangen S (2005) Managing to collaborate: The theory and practice of collaborative advantage. Abingdon: Routledge.
Huyer S (2004) Challenging relations: A labour-NGO coalition to oppose the Canada-US and North American Free Trade Agreements, 1985–1993. Development in Practice 14: 48–60.
Hölstrom M (2006) Globalisation and good work: Impiva, a Spanish project to regenerate industrial districts. Tijdschrift voor Economische en Sociale Geografie 97: 491–502.
Jauhiainen JS (2006) Multipolis: High-technology networks in Northern Finland. European Planning Studies 14: 1407–1428.
Kassler WJ, Goldsberry YP (2005) The New Hampshire public health network: Creating local public health infrastructure through community-driven partnerships. Journal of Public Health Management Practice 11: 150–157.
Kennedy L (1999) Cooperating for survival: Tannery pollution and joint action in the Palar Valley (India). World development 27: 1673–1691.
Kimberly JR (1976) Organizational size and the structuralist perspective: A review, critique, and proposal. Administrative Science Quarterly 21: 571–597.
Klijn EH (2005) Designing and managing networks: Possibilities and limitations for network management. European Political Science 4: 328–339.
Malmborg F von (2007) Stimulating learning and innovation in networks for regional sustainable development: the role of local authorities. Journal of Cleaner Production 15: 1730–1741.
March JG (1991) Exploration and exploitation in organizational learning. Organization Science 2: 71–87.
McKenna VB, Gunn JE, Auerbach J, Brinsfield KH, Dyler S, Barry MA (2003) Local collaborations: Development and implementation of Boston's bioterrorism surveillance system. Journal of Public Health Management Practice 9: 384–393.
Miles RE, Miles G, Snow CC (2005) Collaborative entrepreneurship: How communities of networked firms use continuous innovation to create economic wealth. Stanford, CA: Stanford Business Books.
Moffat L, Archer N (2004) Knowledge management in production alliances. Information Systems and e-Business Management 2: 241–267.
Molina-Morales FX (2005) The territorial agglomeration of firms: A social capital perspective from the Spanish tile industry. Growth and Change 36: 74–99.
Mulrow CD (1994) Systematic reviews: Critical links in the great chain of evidence. Annals of Internal Medicine 126: 389–391.
Navdi K (1999) The cutting edge: Collective efficiency and international competitiveness in Pakistan. Oxford Development Studies 27: 81–107.
Norman C, Axelsson R (2007) Co-operation as a strategy for provision of welfare services: A study of a rehabilitation project in Sweden. European Journal of Public Health 17: 532–536.
Nunan F (1999) Policy network transformation: The implementation of the EC directive on packaging and packaging waste. Public Administration 77: 621–638.
Paniccia I (1998) One, a hundred, thousands of industrial districts: Organizational variety in local networks of small and medium-sized enterprises. Organization Studies 19: 667–699.
Parkinson AJ, Bruce MG, Zulz T and the International Circumpolar Surveillance Steering Committee (2008) International circumpolar surveillance, an Arctic network for surveillance of infectious diseases. Emerging Infectious Diseases 14: 18–24.
Powell WW (1990) Neither market nor hierarchy: Network forms of organization. In: Staw Barry M, Cummings LL (eds), Research in organizational behavior, vol. 12. Greenwich, CT: JAI Press, pp 295–336.

Prahalad CK, Ramaswamy V (2004) The new frontier of experience innovation. MIT Sloan Management Review 45: 12–18.
Provan KG, Fish A, Sydow J (2007) Interorganizational networks at the network level: A review of the empirical literature on whole networks. Journal of Management 33: 479–516.
Provan KG, Kenis P (2008) Modes of network governance: Structure, management, and effectiveness. Journal of Public Administration Research and Theory 18: 229–252.
Provan KG, Milward HB (1995) A preliminary theory of network effectiveness: A comparative study of four community mental health systems. Administrative Science Quarterly 40: 1–33.
Rice J, Juniper J (2003) High technology alliances in uncertain times: The case of Bluetooth Knowledge. Technology and Policy 16: 113–124.
Rogers N, Anderson W (2007) A community development approach to deal with public drug use in Box Hill. Drug and Alcohol Review 26: 87–95.
Rutherford TD, Holmes J (2007) We simply have to do that stuff for our survival: Labour, firm innovation and cluster governance in the Canadian automotive parts industry. Antipode 39: 194–221.
Salancik G (1995) Wanted: A good theory of network organization. Administrative Science Quarterly 40: 345–349.
Salas-Porras A (2005) Changing the bases of political support in Mexico: Pro-business networks and the market reform agenda. Review of International Political Economy 12: 129–154.
Sandelowski M, Docherty S, Emden C (1997) Focus on qualitative methods: Qualitative metasynthesis: Issues and techniques. Research in Nursing and Health 20: 365–371.
Schumaker AM (2002) Interorganizational networks: Using a theoretical model to predict effectiveness of rural health care delivery networks. Journal of Health and Human Services Administration 25: 371–406.
Stern RN (1979) The development of an interorganizational control network: The case of intercollegiate athletics. Administrative Science Quarterly 24: 242–266.
Takahashi LM, Smutny G (2002) Collaborative windows and organizational governance: Exploring the formation and demise of social service partnerships. Nonprofit and Voluntary Sector Quarterly 31: 165–185.
Thompson JD (1967) Organizations in action. New York: McGraw-Hill.
Townsend WA (2004) Systems changes associated with criminal justice treatment networks. Public Administration Review 64: 607–617.
Tranfield D, Denyer D, Smart P (2003) Towards a methodology for developing evidence-informed management knowledge by means of systematic review. British Journal of Management 14: 207–222.
Valadez JJ, Hage J, Vargas W (2005) Understanding the relationship of maternal health behavior change and intervention strategies in a Nicaraguan NGO network. Social Science & Medicine 61: 1356–1368.
Weick KE (1976) Educational organizations as loosely-coupled systems. Administrative Science Quarterly 21: 1–21.
Williamson OE (1973) Markets and hierarchies: Some elementary considerations. American Economic Review 63: 316–325.
Woodward J (1965) Industrial organization: Theory and practice. London, Great-Britain: Oxford University Press.

Part II
Dynamics of Adaptation and Change

Chapter 3
Organizational Trade-Offs and the Dynamics of Adaptation in Permeable Structures

Stephan Billinger and Nils Stieglitz

Abstract Organization design has a critical impact on how firms adapt to the business environment. In our case study, we show how organization design increases a firm's ability to sense and seize business opportunities by making its organizational boundaries more permeable. Our findings reinforce and substantiate prior work on organization design and organizational adaptation. They also suggest how insights from organization design theory may help better understand the dynamic capabilities of firms. We find that disintegration and the creation of a permeable corporate structure require decision-makers to consider four organizational trade-offs: specialization, interdependencies, delegation, and incentives. We discuss how these organizational trade-offs provide a useful complementary perspective to the dynamic capability approach by highlighting the structural properties that shape organizational adaptation across time.

Keywords Organization design · Organizational tradeoffs · Dynamic capabilities · Search · Coordination

3.1 Introduction

Forecasting future market conditions is difficult and imposes considerable demands on a firm's organization design and its ability to adapt dynamically. Organization design must therefore be aimed at enhancing the firm's adaptive capacity to ensure a dynamic fit between environmental conditions and organizational characteristics (Donaldson 1987; Volberda 1996). In strategic management research, the concept of dynamic capabilities has gained currency over the last decade (Eisenhardt and Martin 2000; Teece et al. 1997; Winter 2003). According to this influential perspective, firms need to develop and foster dynamic capabilities to successfully navigate the

S. Billinger (✉)
Strategic Organization Design Group, Department of Marketing and Management, University of Southern Denmark, Odense, Denmark
e-mail: sbi@sam.sdu.dk

A. Bøllingtoft et al. (eds.), *New Approaches to Organization Design*, Information and Organization Design Series 8, DOI 10.1007/978-1-4419-0627-4_3,

demands of changing markets. Dynamic capabilities enable firms to sense and seize business opportunities by continuous realignment of tangible and intangible assets (Teece 2007). Much attention has been paid to the organizational routines and processes that constitute dynamic capabilities. According to Teece et al. (1997: 518), dynamic capabilities are "a learned pattern of collective activity through which the organization systematically generates and modifies its operational routines in pursuit of improved effectiveness." Eisenhardt and Martin (2000) argue that dynamic capabilities are a set of specific and identifiable organizational processes, such as product development, strategic decision making, and managing alliances. This focus on emergent organizational routines and processes has come at the expense of explicitly considering how formal organization design impacts the adaptive capacity of a company. We use a detailed case study to analyze how organization design influences a firm's ability to sense and seize business opportunities.

Organization design theory can inform some key aspects of how firms may improve their dynamic capabilities to adapt to future business opportunities. The ability to sense a business opportunity relates to a firm's ability to search and evaluate its business environment. The behavioral theory of the firm has a long tradition of insisting on organizational search as the main mechanism to adapt to the demands of the task environment (Cyert and March 1963; Grandori 1987; March and Simon 1958; Thompson 1967). Seizing a business opportunity requires the organizational realignment of complementary assets and interdependent activities by integration (Lawrence and Lorsch 1967; Siggelkow and Rivkin 2005; Stieglitz and Heine 2007; Thompson 1967). Thus, how firms sense and seize business opportunities has always been a major concern for organization design theory. This phenomenon can be effectively studied by examining corporations that disintegrated their structure and increased the permeability of their boundaries, that is, their ability to buy and sell from external parties (Jacobides and Billinger 2006). For example, in their detailed case study, Grant and Cibin (1996) show how increased market turbulence in the oil industry led to widespread changes in the organization design of major oil companies (cf. Grant 2003). The structural changes aimed at improving adaptive capacity: "The result was a quest for structure and systems which would be capable of responding quickly to external changes, would foster opportunism [sic] and an entrepreneurial drive for profit, but at the same time would permit planning and investing for long-term development (Grant and Cibin 1996: 180). Specifically, oil majors engaged in vertical disintegration and wide-spread decentralization, restructured their reward systems to foster innovation, and installed financial control systems to motivate and coordinate business units (see also Harrigan 1984). Hence, organization design directly had an impact on the dynamic capabilities of oil majors and how they sensed and seized business opportunities.

In our article, we show how organization design impacted the adaptive capacity of a European apparel manufacturer. Like the oil majors in the 1980s, the firm implemented a disintegrated, more permeable vertical structure. Our case study shows how making a value chain more permeable, that is, opening op the boundaries toward external markets, led to major changes in the organization design. These changes confront the organization designer with organizational trade-offs that turn out to be the structural design mechanism for organizational disintegration. We illustrate how the organizational trade-offs of specialization and interdependencies determine task-related mechanisms, and the trade-offs of delegation and incentives define people-related mechanisms. We find that balancing all four organizational trade-offs is critical for organizational adaption and for building and sustaining dynamic capabilities. Moreover, our findings suggest that increasing the adaptive capacity of a company comes at the price of more complex coordination and integration. We also observe that coordination, in turn, influences how firms search for business opportunities. We find that both organizational search and coordination are limited by organizational designer's ability to adjust the organizational trade-offs within the disintegrated permeable structure.

This chapter is structured as follows: Firstly, the methods section will describe the setting, data collection and analysis. We also briefly discuss why and how the firm disintegrated its structure. We then turn to the case analysis, in which we discuss how the firm changed its patterns of sensing and seizing, and how the organizational structure facilitated this change. Finally, we discuss findings and implications.

3.2 Methods

3.2.1 Setting

We studied Fashion Inc., a major European apparel manufacturer, which disintegrated its value chain between 2003 and 2005. Before 2003, the firm had a traditional vertically integrated structure: Fashion Inc. purchased fiber from external sources and then manufactured complete apparel collections, which it then sold to retailers. All these steps took place in Western Europe. Due to changes in the retail structure (i.e., fewer independent local retailers, emerging low-cost retail chains) and the emergence of low-cost production countries (e.g., Asia), Fashion Inc. had to abandon this traditional business model. What makes this case interesting is that Fashion Inc., rather than outsourcing its European production, "opened up" its value chain in the various stages of production, and it created a permeable vertical architecture (Jacobides and Billinger 2006) with three distinct Strategic Business Units (SBUs). Fashion Inc. started selling to and purchasing from external parties at various steps of the value chain: For instance, the Fabric Unit engaged in selling fabric to other apparel manufacturers; the CMT (Cut Make Trim) Unit offered free production capacities to external manufacturers, and the OBM (Original Brand Name) Unit offered design and logistics to firms that only engaged in branding (see also Fig. 3.1).

With the implementation of this new business model, Fashion Inc. also had to redesign its traditional organizational structure, similar to the changes made by oil companies in the 1980s (Harrigan 1984; Grant 2003). What is new in this case is the variety of operational modes and strategic opportunities that this new structure facilitated. To understand this newly established dynamic capability, this article

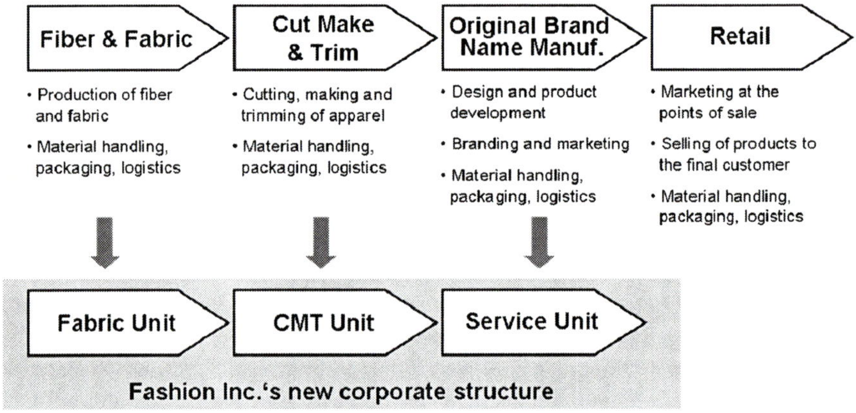

Fig. 3.1 The apparel value chain and Fashion Inc.'s new corporate structure

concentrates on the organization design choices made by Fashion Inc. and what effect those choices had on the firm's ability to sense and seize opportunities.

3.2.2 Data Gathering

Between 2003 and 2005 we were fortunate enough to be able to observe more than 200 reengineering workshops within the firm. We also received access to Fashion Inc.'s internal archives and documentation of the organizational design choices. Moreover, we conducted more than 120 interviews within the firm. We thoroughly examined all these data sources in order to identify how the firm changed its sense and seizing patterns. We did so by looking at the firm's specialization patterns within the SBUs and the interdependencies between organizational units. We then looked at the freedom to accomplish tasks (delegation) for organizational actors, as well as incentives created to motivate them. In particular, we found useful information in the business process mapping of the firm, as well as in its newly-established job descriptions. Through interviews, we confirmed our findings and refined them wherever necessary. The entire research process followed basic principles of qualitative research (Eisenhardt and Graebner 2007; Miles and Huberman 1994; Yin 2003). The summary of the data sources and the research process can be found in Table 3.1.

3.2.3 Data Analysis

For the data analysis we used theoretical constructs that are well-supported in various streams of literature: We first examined *specialization*, which is the decomposition and assignment of tasks to organizational members (Becker and Murphy

Table 3.1 Sources of evidence throughout the study

Sources of evidence in each stage of the study	Stage 1: June 2002–January 2003	Stage 2: January 2003–February 2004	Stage 3: February 2004–May 2006
Primary sources of data	· Workshop participation, workshop documentation (e.g., handouts, workshop transcripts, working documents, process maps) · Project management documentation · Personal research notes · Internal documents · SBU business plans · Ongoing discussions with project management team	· Workshop participation, workshop documentation (e.g., handouts, workshop transcripts, working documents, process maps) · Documentation for IT requirements · Project management documentation · Internal documents · Personal research notes · Employee survey · Ongoing discussions with project management team	· Workshop participation, workshop documentation (e.g., handouts, workshop transcripts, working documents, process maps) · Internal documents · Personal research notes · Project management documentation · IT-design documents · Ongoing discussions with project management team · Semi-structured interviews to confirm theory-building
Secondary sources of data	· Historical studies of Fashion Inc. · Sector descriptions · Research papers with apparel focus · Analyst reports	· Sector descriptions · Press releases · IT-manuals · Company manuals	· Sector descriptions · Press releases · IT-manuals · Company manuals
Company events	· BPR workshops · Firm-wide gatherings (1 presentation of the new collection, 1 firm anniversary, 2 firm parties	· BPR workshops · Firm-wide gatherings (1 presentation of the new collection, 2 firm parties)	· BPR workshops · Firm-wide gatherings (1 presentation of the new collection, 1 firm party)

1992; Blau 1972; March and Simon 1958; Simon 1947). Then we studied *interdependencies* between subtasks and performance outcomes of these interactions (Harris and Raviv 2002; Levinthal 1997; Thompson 1967). We continued by examining *delegation*, that is, constraints placed on how organizational members accomplish a subtask (Mintzberg 1979; Milgrom 1988); and, lastly, we studied *incentives* as the rewards for the organizational members' performance within a subtask (Kaplan and Henderson 2005). These four aspects reflect traditional dimensions of organization design, and they have individually, or in combination, been used in various research settings (see references above). They provide a useful analytical lens for studying Fashion Inc.'s disintegration, in particular the reengineering workshop protocols, job descriptions and business process handbooks.

3.3 Disintegrating Traditional Vertical Integration: Why Fashion Inc. Disintegrated

In the late 1990s, Fashion Inc. was coming under severe pressure from its end market. The firm's traditional strategy focused on the design, manufacture and distribution of Fashion Inc.'s own brand, which targeted the mid-price market segment. This business model was also used by most of Fashion Inc.'s competitors, creating an industry constellation in which comparable firms were competing for similar market segments. Consequently, prices were falling and all players were forced to review their strategies. The situation was exacerbated because overall European spending on apparel was decreasing. In addition, consumers' behavior was changing, and the mid-price market segment, Fashion Inc.'s main market, was losing market share to Asian low-cost imports and global brands (for a discussion, see the analysis of ZARA and H&M by Deutsche-Bank 2002).

To identify its future strategy, Fashion Inc. analyzed the entire firm and explored the potentials along its existing value chain. It found that the firm's profit could be derived from all activities of its value chain, and that the industry's bandwagon behavior of "concentration on branding" meant that these downstream activities were becoming more competitive. A senior manager summarized this view by saying:

> Everyone seems to do the same thing: Getting rid of manufacturing and focusing on the end customer. It seems that they all want to become ZARA or H&M – and they forget that in particular these two firms still have own manufacturing... Also, many of those who have outsourced are everything but happy about their decision, as products from Asia oftentimes do not have the quality that these firms want...

Following this mindset, it became apparent that other parts of the value chain could be more attractive. Fashion Inc. concluded that its competitive strengths were also in upstream activities such as design, upstream production, logistics, and fabric development. A manager of the marketing department concluded: "Many competitors outsource – why shouldn't we help them?"

Fashion Inc.'s senior management reached the conclusion that disintegrating the value chain was a promising option, that is, the firm would establish distinct SBUs along its value chain: The Fabric Unit, the CMT Unit (cutting, making, and trimming) and the Service Unit (see Fig. 3.1). This would expose the firm to various (intermediate) markets and allow the firm to sense and seize additional business opportunities, as one manager outlined:

> It is really amazing that there are only few firms operating in some of the upstream markets. It seems to me that they not only have healthy margins but also have no problems finding customers for the products they produce.

Fashion Inc.'s notion was therefore that the new SBUs would be able to sell to internal units as well as to external customers and they would be allowed to buy from internal and external suppliers. In other words, Fashion Inc. decided to make its value chain more permeable to sense and seize business opportunities in intermediate markets.

The strategic rationale for disaggregating the corporate structure was rooted in three types of benefits, which explained why Fashion Inc. did not change its overall vertical scope. The three benefits are: First, direct interactions with intermediate markets allowed the firm to search for new business opportunities along the entire value chain and to develop new strategic capabilities (e.g., advanced fabric development). Second, the exposure to intermediate markets enabled the firm to benchmark its internal resources against external competitors and to use this information for corporate resource allocation. Third, operational efficiency would increase as a result of standardized processes within the firm's new modular structure. The strategic logic of these three benefits is discussed in detail in Jacobides and Billinger (2006), who argue that this rationale is the basis for Fashion Inc.'s permeable vertical architecture as well as the overall boundary design and vertical scope of the disintegrated structure.

3.4 Case Analysis

3.4.1 Sensing and Seizing Opportunities in the Old and New Structure

In the old vertically integrated structure, Fashion Inc. executed business in a way that is comparable to traditional examples of vertical integration. Fashion Inc. produced apparel collections under its own brand for retail stores in Europe. The firm was primarily concerned with sensing future business opportunities for the established market segments that it already operated in. A former sales agent summarized this by saying:

> Traditionally our established brand was very strong. We had very high market shares in several market segments and whatever we produced the retail stores would sell. There was also not much change in the market, and we could easily keep on doing what we always did... In addition, our products were oftentimes standard, high-quality products that did not

change much; they simply followed the seasonal cycles. We therefore had stable production, and manufacturing could plan with fairly accurate forecastings.

The statement illustrates that Fashion Inc. did not sense and seize many opportunities and that the original business model was primarily based on stable production. Whenever the firm identified new opportunities, the firm also invested significant time and money in realizing them, as one product manager said:

Whenever we identified a new product segment or we wanted to produce a new collection, we first searched for the right fibre, and then we tested the knitting and the fabric. Finally we produced prototypes that several people assessed. Oftentimes changes were made, affecting one or all of the various steps. This took a lot of time, in particular because we also wanted to produce superior quality.

This approach of new product development was not only costly but also led to a situation in which Fashion Inc. could not generate sufficient revenue from a high-quality innovation. The firm therefore had to identify new markets, and it chose the various intermediate markets along the value chain. A senior manager developed a vision that nicely circumscribed the organizational design paradigm with which the firm wanted to realize its strategic intention:

If we want to serve a customer we want to be able to deliver the same standards in terms of delivery time and quality, options available and so on... with standardized processes we can achieve this easily, regardless of the actual product... whether the customer purchases fabric or a t-shirt or something else, a modular 'order processing' allows to plug and play for different products in different SBUs.

One example that shows how Fashion Inc. realized this intention was the licensing of other fashion brands. This approach allowed Fashion Inc. not only to better leverage existing innovations but also to address more market segments by being able to differently position the various brands. A manager pointed out:

When you compare many of the existing apparel brands, also in the high-price segment, you will not find many differences. Fabric is oftentimes identical, there may only be a different colour or pattern, and even the gadgets have similar features. Having a modular LEGO-like portfolio with product features obviously allows you to produce products for many different brands.

What is remarkable for this statement is that several mangers along Fashion Inc.'s value chain argued this way, that is, the firm found a similar situation in fabric production or in the value adding stage of "cutting, making and trimming." Managers identified that not solely concentrating on one brand, but being able to produce products for multiple brands could be a lucrative proposition. A manager stated:

Multi-brand strategies are very common in other industries, and also when you have a look at the Asian apparel manufacturers: they produce for many different, oftentimes competing, firms. Why should we not do the same?

The above illustrations indicate that Fashion Inc., while being under pressure with its own brand, had a number of opportunities along its value chain. Opening up its boundaries and seizing these opportunities was therefore a reasonable choice, as it allowed the firm to directly engage with new customers *without* making major investments, such as the ones that a traditional differentiation strategy would typically require, as Fashion Inc.'s approach considered opportunities along the vertical dimension of the value system, rather than the horizontal dimension. Moreover, Fashion Inc.'s approach also considers both buying from and selling to external parties at several stages of the value chain, which is also different from the oil majors in the 1980s, which mostly established business around one business unit that engaged in refining oil and associated services (see Harrigan 1984), and did not engage in multiple forms of business, as did Fashion Inc. with its three distinct SBUs and a distinct strategic rationale for its permeable architecture and vertical scope (Jacobides and Billinger 2006).

The above quotes already indicate that organization design choices were critical for Fashion Inc. to develop a structure with dynamic capabilities. We now briefly summarize how Fashion Inc. implemented the disintegrated structure using a reengineering team, and then turn to the organization design choices necessary to implement the structure. We ask (a) how did Fashion Inc. define tasks (specialization) and interdependencies, and (b) how did it reward people (incentives) and delineate their freedom to act (delegation)?

3.4.2 Specialization and Interdependencies Within a Disintegrated Structure

Fashion Inc. defined specialization and interdependencies by using its business processes as the central unit of analysis; and, to be more precise, the firm did so by examining its functional activities to delineate tasks and activities. Within each SBU these were obvious tasks, such as knitting and dying in the Fabric Unit, cutting and sewing in the CMT Unit, and sales and retail marketing in the OBM Unit. A manager remarked:

> Within the new business model we do not produce any new products: It is still fabric or t-shirts or whole collections. What changes is simply [the] different type of customers that we now have [along the value chain].

Fashion Inc. summarized all of its activities with an overall business process framework which can be found in Fig. 3.2.

The business process framework allowed Fashion Inc. to demarcate individual activities (specialization) and standardize outputs (interdependencies) while making them usable for the next value-adding step – regardless of whether this step was internal or external. Interesting to note is that, within each SBU, Fashion Inc. created business processes that were partially specific for the SBU and partially generic within the organization. For instance, each SBU had its own product development that was distinct from the other SBUS; and it had a generic order processing that all SBUs used in the same way. As a result, Fashion Inc. not only disaggregated *along* the value chain into three distinct SBUs, but also "unbundled" within each step of the value chain, that is, it split up activities within an SBU into customer relationship

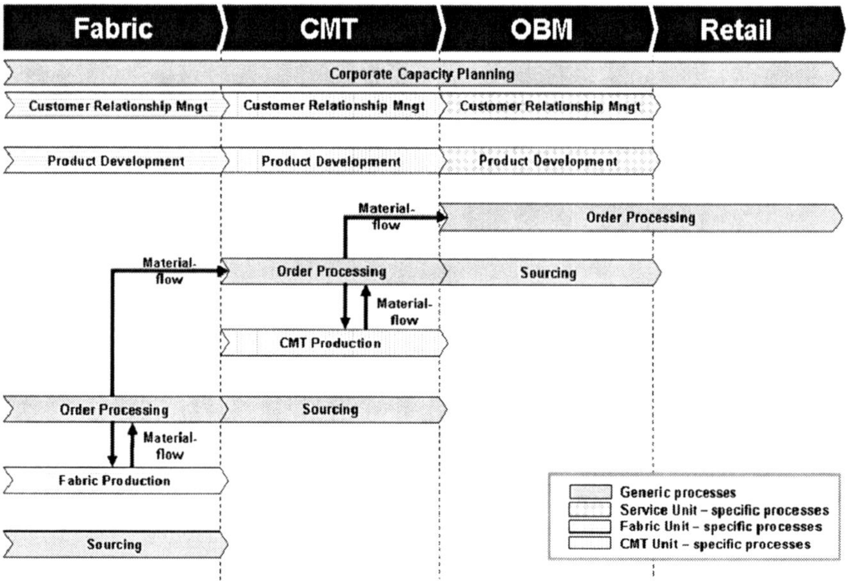

Fig. 3.2 Fashion Inc.'s overall process architecture

management, product development, order processing, and fabric production (for a discussion of the latter, see Hagel and Singer 1999). The firm thereby created a "LEGO-like" (see quote above) process architecture as a central backbone for the newly established disintegrated structure.

3.4.3 Delegation and Incentives Within a Disintegrated Structure

Delegation and incentives were required for Fashion Inc. to assign tasks to individual people and to create adequate rewards. For delegation, or the freedom to accomplish tasks, Fashion Inc. redesigned the job descriptions for many employees. It did so by clearly allocating individuals to specific roles and business processes. For instance, in the area of product development Fashion Inc. had to conduct Trend Research, and the firm allocated particular roles for each individual by clarifying (a) who had the responsibility for the task, (b) who was executing it, (c) who was participating in it, and (d) who gave input. Figure 3.3 briefly illustrates how this delegation was executed. Another example was the delegation of more decision-rights to the CEOs of the SBUs, who had to use their new autonomy (i.e., "looser" delegation) to freely identify new markets.

A quote from a manager illustrates the rationale for delegation:

			Product DevelopmentDivision				
Process	**Task**	**Department** **Process in process-handbook**	**Product Manage-ment**		**Design**		**...**
Employee			*Ann*	*Peter*	*Joe*	..	
Research Develop ment &	Color Development	1.7.2 R&D– Worshop	R	I	P	...	
		1.2.1 Trend Research	P	R	-	...	...
		1.5 3 Color Scheme	-	E	R	..	...

Key:

I	Input	Employee gives Input
P	Participation	Employee is participating
E	Execution	Employee is executing
R	Responsibility	Employee has the responsibility

Fig. 3.3 Allocation of decision and use rights (we use the fake names Ann, Peter and Joe; the spreadsheet itself stems directly from the reengineering workshop)

> Having a process diagram and business process handbook is nice, but it does not help you implement the business model. If you want this new system [i.e., the disintegrated structure] to work, you need to tell every single person to do what and when.

On the basis of delegation, Fashion Inc. then created incentives for individuals. For instance, middle managers responsible for logistics within an SBU were rewarded on a set of operational performance measures. These key performance indicators were then summarized with a Balanced Scorecard, which allowed a systematic performance monitoring within the disintegrated structure. Fashion Inc. largely followed principles similar to those described by Kaplan and Norton (1996).

To better understand the interplay of organizational design choices, and how they affected search and coordination within Fashion Inc, we now turn to two specific examples that illustrate in detail how the firm's new organizational design influenced organizational adaptation.

3.4.4 How the Disintegrated Structure Solved Major Challenges of Traditional Vertical Integration

Fashion Inc.'s original vertically integrated structure had a rigid functional organizational form with Sales and Marketing, Production, Logistics and Product Development departments. Senior management coordinated all these departments and ensured efficient operations. This could only be guaranteed with a stable production regime that strictly followed seasonal cycles. Any irregularities caused major additional coordination costs which the organization could not bear in most instances.

Major changes in the market, including shorter fashion cycles, more differentiated products as well as increased volatility in demand caused numerous challenges within the old structure. This can be traced down to Fashion Inc.'s internal organizational design, that is, its specialization, interdependencies, delegation, and incentives within the old structure. Specialization followed functional necessities for the *overall* corporate structure. Interdependencies were purely designed to have stable and efficient production without changing customer demand and without considering intermediate markets (see also quotes from above). Delegation, that is, the freedom to accomplish a task, was driven by mutual adjustment and senior management intervention, and did not allow for local adaptation. Finally, incentives were all designed to only optimize global corporate performance – and did not consider opportunities along the firm's value chain. Addressing the challenges and changing the organizational design was obviously a tedious and painful process (as also illustrated by the BPR literature). To provide an example of how Fashion Inc. identified challenges and addressed them during the BPR, the following instance discusses the evolution of incentives and delegation on the SBU level.

Fashion Inc.'s business planning for the new structure assigned full autonomy (delegation) for SBU managers and considered rewards that were 100% based on the SBU's performance (incentives). However, as it soon turned out, this arrangement jeopardized some of the overall corporate goals and undermined coordination between SBUs. For instance, managers from different SBUs started to engage in heavy negotiations and they were less willing to cooperate with other SBUs to deliver jointly developed products. One manager said: "If we are independent, why should we cooperate with the others – we would be better off without the others."

Corporate management's view, however, was that this behavior would not allow the firm to generate firm-level benefits such as rapid response products (see below). As a result of the full autonomy with 100% SBU reward system, senior executives were dragged into mediating the negotiations between SBUs, which used up valuable management time and created an unwanted situation for the firm as a whole. Fashion Inc. therefore implemented an adjusted reward system in which managers were rewarded based on corporate results (50%) and SBU performance (50%). This compromise ensured that SBU managers considered the goals of both the individual business unit and the corporation as a whole and that lengthy negotiations between SBUs were limited. The CEO pointed out:

> We do not have these major discussions any more ... and people now acknowledge each other's turf and also see the whole..." He added in another instance: "The 50:50 rule also ensures that a CEO of a SBU gets the benefits of his efforts."

The arrangement also ensured that global (corporate) and local (SBU) efforts to identify new business opportunities would not completely jeopardize each others' objectives, as autonomy was still granted but intra-organizational adaptation was simultaneously fostered. This example illustrates that the constraints placed on how organizational members accomplish subtasks (i.e., delegation or the level of autonomy), can be characterized as either loose or rigid. And incentives, that is, the rewards that organizational members receive for accomplishing assigned tasks, can be characterized as either individual or collective. In the previous structure, Fashion Inc. had loose constraints as individual incentives for middle managers solely based on the overall firm's performance (which allowed for free-riding). In the new disintegrated structure, a mixture of collective and individual performance incentives was implemented to mediate the autonomy. Fashion Inc.'s senior management realized that the SBUs' efforts to sense and seize opportunities in intermediate markets must be motivated by high-powered incentives for the SBU managers. On the other hand, these SBU incentives and full autonomy undermined firm-wide coordination and jeopardized corporate capabilities. The compromise implemented – a mixture of individual and collective rewards that controlled the autonomy of individual SBUs – reflects an example in which "sensing and seizing" business opportunities required a fine balance of the trade-offs associated with incentives and delegation.

3.4.5 Fashion Inc.'s New Organizational Design in Action: The Case of Rapid Response

Another example that illustrates how Fashion Inc.'s new organizational structure allowed for new ways to sense and seize business opportunities was the firm's new "rapid response" program (for a discussion see Ghemawat and Nueno 2003; or Richardson 1996). This program was designed to address a new emerging customer segment which required Fashion Inc. to offer fashionable products quickly. Accommodating these products was too time-consuming with negotiations between the SBUs, and Fashion Inc. therefore had to find a coordinated joint rapid response activity which was similar to a traditional vertical integration. However, allowing Fashion Inc.'s corporate planning department to plan the SBUs' resources was a tricky issue. One central planning department manager explained:

> If you want to have a quick response program, you have to be able to plan the entire [value] chain... for instance, for some products it might make sense to have high inventory levels in the fabric warehouse and in other cases you have high inventory levels for finished products... with centralized planning, we can optimize these different requirements... how can an SBU do that?

Hence, this manager was proclaiming that a broad task assignment for the central planning department was advantageous in facilitating SBU coordination. In contrast, an SBU manager replied:

> ...if we are supposed to address our own [SBU] markets, we cannot have a centralized planning department that constantly tells us what we can do and what we cannot do ...

This manager was suggesting that decentralized efforts to find business opportunities (within the SBU) should not be undermined and hampered by more centralized coordination. After several discussions (and some experimentation) it became clear that the central planning department needed only to direct specific manufacturing resources whenever prospective rapid response products were identified by the Service Unit. A compromise was reached: whenever these resources were not being used as directed by the central planning department, the SBU could use them for its own purposes, that is, opportunities that it discovered from its own search activities in the SBU's (intermediate) markets.

The above example illustrates that specialization and interdependencies had a critical impact upon how Fashion Inc. sensed and seized business opportunities, in particular in complex situations such as the rapid response program which required a high degree of inter SBU-integration. Fashion Inc. essentially had to decide between a broad and a more narrow assignment of tasks for the central planning department as well as weak or strong interdependencies between units. In the former structure, the central planning department had a broad task assignment with strong interdependencies. The department was responsible for the coordination of all of Fashion Inc.'s resources and for ensuring that end-market search was effectively supported. In the new disintegrated structure, the task assignment for the central planning department was narrower and the interdependencies were weaker, as many planning tasks were assigned to the SBUs. By balancing the trade-offs associated with specialization and interdependencies, Fashion Inc. was able to create a functioning disintegrated structure.

3.5 Discussion and Implications

The case study provides insight on how organization design affects a firm's ability to adapt to its environments. Organizational adaptation requires both the ability to sense and to seize business opportunities. Sensing business opportunities fundamentally relates to how firms search and evaluate business opportunities. Fashion Inc. strengthened its sensing ability by disintegrating its vertical structure and increasing its permeability, allowing business units to search external business opportunities in intermediate markets. Disintegration entailed the delegation of decision rights by placing far less organizational constraints on SBU managers. Fashion Inc.'s senior management motivated the search for business opportunities by providing more high-powered incentives for SBU managers. Seizing business opportunities for the entire organization, on the other hand, requires coordination across SBUs. This is especially true for the company-wide rapid response program that requires the coordinated effort of multiple SBUs, but which also hampered the SBUs' freedom. To facilitate coordination across SBUs, senior management decided to standardize functional interfaces, transforming reciprocal interdependencies into sequential interdependencies (Thompson 1967). This greatly simplified coordination across SBUs and enabled a more rapid response to business opportunities. Processes within SBUs were also more heavily standardized and clearly defined. In addition, rewards based only on SBU performance, according to senior management, undermined the firm's ability to coordinate the SBUs. Senior management therefore paddled back a bit by introducing incentives that depend on both corporate and SBU performance. Senior management's hope was that this would help to bring about coordination across SBUs. Overall, these changes in organization design strengthened the dynamic capabilities of Fashion Inc. and translated into an increased corporate ability to adapt to the multiple business environments along the value chain. This suggests that formal organization design is clearly important for understanding how firms attain dynamic capabilities. Our case shows that, in particular, organizational trade-offs are a fundamental pillar for the organization design of disintegrated permeable structures. They provide a useful complementary perspective to the dynamic capability approach by highlighting the structural properties that shape organizational adaptation across time.

Furthermore, our case study also suggests that Fashion Inc. had to make some hard choices in designing the new organizational structure. The senior management was confronted by organizational trade-offs relating to the firm's ability to sense and seize business opportunities. For example, in setting up the incentive system, senior management had to balance the conflicting demands of motivating more intensive search for business opportunities and of more effort at coordination across SBUs (cf. Lorsch and Allen 1973). Designing the interdependencies among SBUs also led to hard choices. Allowing for strong interdependencies would have increased corporate ability to sense business opportunities along several steps of the value chain, but only at the cost of more intensive coordination (cf. Thompson 1967; Harris and Raviv 2002). Specialization makes internal coordination more demanding (Lawrence and Lorsch 1967; Blau 1972), but improves the ability to sense opportunities, since a specialist can focus on a narrower domain of the environment. Fewer constraints on delegated decisions facilitate experimentation and search, thereby increasing the firm's ability to sense business opportunities. However, it also calls for more intensive coordination by feedback, because the activities of subordinates tend to be less standardized and predictable. We have collected possible propositions suggested by the case study and by prior research in Table 3.2.

A potential reason for these suggested organizational trade-offs might be that increasing the scope and intensity of search for business opportunities within a task increases the variance in the alternatives (Denrell 2003; Denrell and March 2001). This increases the demand for coordination by feedback (March and Simon 1958), since a greater variance implies less room for coordination by plan. Search and coordination by feedback thus complement each other. However, coordination by feedback tends to be more costly, since agents need to allocate considerable time to aligning their activities. This not only cuts down the time agents may spend on searching and evaluating business opportunities but also slows down their ability to rapidly seize business opportunities. While our case study provides evidence for the presence of these important organizational trade-offs, we also view this as a promising area for further research.

Summing up our contribution, our insights further develop existing research in various ways: First, we identify the organizational trade-offs that organizational designers are confronted with when firms simultaneously engage in multiple sensing and seizing processes. Second, we show that these trade-offs are fundamental

Table 3.2 Organizational trade-offs in the coordination of search processes (stylized)

Dimension	Definition	Trade-off	Search	Coordination
Specialization	Decomposition and assignment of tasks to organizational members	Narrow – broad	Narrow = search ↑ Broad = search ↓	Narrow = coordination ↓ Broad = coordination ↑
Inter-dependencies	Interactions between tasks that result in performance outcomes	Weak – strong	Weak = global adaptation ↓ Strong = global adaptation ↑	Weak = coordination ↑ Strong = coordination ↓
Delegation	Constraints placed on how organizational members accomplish a task	Loose – rigid	Loose = search ↑ Rigid = search ↓	Loose = coordination ↓ Rigid = coordination ↑
Incentives	Rewards for the organizational members' performance within a task	Individual – collective	Individual = search ↑ Collective = search ↓	Individual = coordination ↓ Collective = coordination ↑

↓: Impedes ↑: Facilitates

for our understanding of organizational design within disintegrated permeable structures. Third, our findings provide a new lens by which organization design adds a structural dimension to research on dynamic capabilities. Fourth, the implications highlight important mechanisms within organizational adaption that could, in particular, enrich the formal modeling of adaption processes. With these insights we hope to enrich theoretical debates.

3.6 Concluding Remarks and Managerial Implications

The grounded approach of our research design provided us with an understanding of how organizational design influences a firm's ability to adapt. An interesting aspect was the observation that many of the managerial change methods employed were well established within the firm. For instance, the organization of Fashion Inc.'s transformation process relied on project management and standardization relied on business process reengineering. Within this context, it was remarkable that some very fundamental organizational aspects were not explicitly covered, or were even partially neglected, during the process. Moreover, our analysis and the comparison with "theory in use" suggest that the balance of the four organizational trade-offs is not adequately addressed by any of the existing toolboxes. We therefore suggest that the four organizational trade-offs – specialization, interdependencies, delegation, and incentives – could be used as guiding principles of organization design, as they allow, in particular, disintegrated permeable firms to analyze organizational structure and explicate organizational designers' choices. This, in turn, also raises important aspects for the theoretical debate, as our case analysis indicates that the discussion of search and coordination *without* the consideration of organizational trade-offs fails to take account of the organizational embeddedness of search processes. With our analysis, we hope to contribute to research on organization design and how it influences the coordination of search processes within and between organizations.

References

Becker GS, Murphy KM (1992) The division of labor, coordination costs, and knowledge. Quarterly Journal of Economics 107 (4): 1137.

Blau PM (1972) Interdependence and hierarchy in organizations. Social Science Research 1 (1): 1–24.

Cyert RM, March JG (1963) A behavioral theory of the firm. Englewood Cliffs: Prentice Hall.

Denrell J (2003) Vicarious learning, undersampling of failure, and the myths of management. Organization Science 14 (3): 227–243.

Denrell J, March JG (2001) Adaptation as information restriction: The hot stove effect. Organization Science 12 (5): 523–538.

Deutsche-Bank (2002) H&M & Inditex – focus on the figures (and not just the fashion). DeutscheBank: Frankfurt am Main.

Donaldson L (1987) Strategy and structural adjustment to regain fit and performance in defence of contingency theory. Journal of Management Studies 24 (1): 1–24.

Eisenhardt KM, Graebner ME (2007) Theory building from cases: opportunities and challenges. Academy of Management Journal 50 (1): 25–32.
Eisenhardt KM, Martin JA (2000) Dynamic capabilities: What are they? Strategic Management Journal 21 (10/11): 1105–1121.
Ghemawat P, Nueno JL (2003) Zara: Fast fashion. Harvard Business School Case # 9-703-497: 1–35.
Grandori A (1987) Perspectives on Organization Theory. Ballinger Publishing Company: Cambridge, MA.
Grant RM, Cibin R (1996) Strategy, structure and market turbulence: The international oil majors, 1970–1991. Scandinavian Journal of Management 12 (2): 165–188.
Grant RM (2003) Strategic planning in a turbulent environment: Evidence from the oil majors. Strategic Management Journal 24(6): 491–517.
Hagel J, Singer M (1999) Unbundling the Corporation. Harvard Business Review 77 (2): 133–141.
Harrigan KR (1984) Formulating vertical integration strategies. Academy of Management Review 9 (4): 638–652.
Harris M, Raviv A (2002) Organization Design. Management Science 48 (7): 852–865.
Jacobides MG, Billinger S (2006) Designing the boundaries of the firm: From "make, buy or ally" to the dynamic benefits of vertical architecture. Organization Science 17 (2): 249–261.
Kaplan S, Henderson R (2005) Inertia and incentives: Bridging organizational economics and organizational theory. Organization Science 16 (5): 509–521.
Kaplan RS, Norton DP (1996) Using the balanced scorecard as a strategic management system. Harvard Business Review 74 (1): 75–85.
Lawrence PR, Lorsch JW (1967) Organizations and environment: Managing differentiation and integration. Harvard University Press: Boston MA.
Levinthal DA (1997) Adaptation on rugged landscapes. Management Science 43 (7): 934–951.
Lorsch JW, Allen SAI (1973) Managing diversity and interdependence. Harvard University, Division of Research, Graduate School of Business Administration: Boston.
March JG, Simon HA (1958) Organizations. John Wiley: New York.
Miles MB, Huberman AM (1994) Qualitative data analysis, 2nd edn. Sage: Thousand Oaks, CA.
Milgrom P (1988) Employment contracts, influence activities, and efficient organization design. Journal of Political Economy 96(1): 42–60.
Mintzberg H (1979) The structuring of organizations. Prentice-Hall: Englewood Cliffs, NJ.
Richardson J (1996) Vertical integration and rapid response in fashion apparel. Organization Science 7 (4): 400–412.
Siggelkow N, Rivkin JW (2005) Speed and search: Designing organizations for turbulence and complexity. Organization Science 16 (2): 101–122.
Simon HA (1947) Administrative Behavior. MacMillan: New York.
Stieglitz N, Heine K (2007) Innovations and the role of complementarities in a strategic theory of the firm. Strategic Management Journal 28 (1): 1–15.
Teece DJ (2007) Explicating dynamic capabilities: The nature and microfoundations of (sustainable) enterprise performance. Strategic Management Journal 28 (13): 1319–1350.
Teece DJ, Pisano G, Shuen A (1997) Dynamic capabilities and strategic management. Strategic Management Journal 18 (7): 509–533.
Thompson JD (1967) Organizations in action. McGraw-Hill: New York.
Volberda HW (1996) Toward the Flexible Form: How to Remain Vital in Hypercompetitive Environments. Organization Science 7 (4): 359–374.
Winter SG (2003) Understanding dynamic capabilities. Strategic Management Journal 24 (10): 991–995.
Yin RK (2003) Case study research: Design and methods, 3rd edn. Sage Publications: Thousand Oaks, London, New Delhi.

Chapter 4
Unpacking Dynamic Capability: A Design Perspective

Deborah E. M. Mulders and A. Georges L. Romme

Abstract This chapter[1] reviews the dynamic capability literature to explore relationships between definition, operationalization, and measurement of dynamic capability. Subsequently, we develop a design-oriented approach toward dynamic capability that distinguishes between design rules, recurrent patterns of behavior, operating routines and processes, market and competitive conditions, and performance outcomes. This framework serves to develop a number of propositions for further research. As such, we integrate the literature on dynamic capability that primarily draws on economics, with a design-oriented approach.

Keywords Dynamic capability · Organizational dynamics · Organizational Adaptation · Organizational design · Design rules · Recurrent patterns of behavior

4.1 Introduction

Today's business environments are fast-moving and open to global competition. This implies that firms need to develop major capabilities in managing change (Teece 2007; Verona and Ravasi 2003). Therefore, a fundamental question in the field of organization studies is how firms can develop the capacity to become more responsive to changes in market and competitive conditions (e.g., Henderson and Clark 1990; Kogut and Zander 1992; Teece et al. 1997). An increasing number of scholars promote the notion of dynamic capability to explain how certain firms achieve competitive advantage in situations of rapid and unpredictable change (Eisenhardt and Martin 2000; Priem and Butler 2001; Sirmon et al. 2007; Teece

D.E.M. Mulders (✉)
Eindhoven University of Technology, Eindhoven, The Netherlands
e-mail: d.e.m.mulders@tue.nl

[1]We thank participants of the 23rd EGOS Colloquium and reviewers of our work submitted to the Academy of Management 2008 Annual Meeting for useful comments and suggestions. A special thanks to Samina Karim and editor Charles Snow, and to the other participants of The Third International Workshop on Organization Design, for their helpful comments.

A. Bøllingtoft et al. (eds.), *New Approaches to Organization Design*, Information and Organization Design Series 8, DOI 10.1007/978-1-4419-0627-4_4,

et al. 1997). The dynamic capability view (DCV) focuses on the dynamic processes of generating, developing, and accumulating a firm's resources, that is, the physical, human, and organizational inputs into the firm's value chain (Eisenhardt and Martin 2000; Javidan 1998; Teece et al. 1997). The DCV stresses the importance of path dependency, the history of a firm's current capabilities (i.e., to exploit its resources), as well as the importance of revising and reconfiguring these capabilities in the future as a response to market changes, and as such achieve and sustain a competitive advantage (e.g., Eisenhardt and Martin 2000; Javidan 1998; Teece et al. 1997; Zollo and Winter 2002). In this chapter, dynamic capabilities are defined as deliberate knowledge that is repeatedly applied in changing operating routines and processes.

The concept of dynamic capability has been predominantly subject to a theoretical debate (e.g., Eisenhardt and Martin 2000; Teece et al. 1997; Zollo and Winter 2002) and empirical research is still rare and exploratory in nature (e.g., Prieto and Easterby-Smith 2006; Sher and Lee 2004; Verona and Ravasi 2003). A fundamental challenge is to develop measures of dynamic capability that are grounded in existing theory, are empirically straightforward and valid (i.e., do not include direct or indirect measurements of firm performance, or specific rules and behaviors, that can raise a tautology problem), and serve to help practitioners make their organizations more effective in revising and reconfiguring their capabilities to exploit its resources. In order to respond to this challenge, we investigate the dynamic capability literature to explore relationships between definition, operationalization, and measurement of dynamic capability. As such, we assess what the collective understanding of dynamic capability appears to be at this point in time. By sampling a large number and broad range of papers, rather than focusing on the consensus list of key papers, this study differs from prior reviews of the dynamic capability literature (e.g., Wang and Ahmed 2007; Schreyögg and Kliesch-Eberl 2007).

Subsequently, we develop an integrated model of dynamic capability that draws on a design-oriented approach. In this respect, the notion design rules (representing purposeful and thoughtful (emerging or established) mental models of people) serves to unpack and operationalize what constitutes dynamic capability, in relation to how it is exposed in recurrent patterns of behavior (referring to a firm's organizational and strategic processes over time), as well as firm performance. Design rules thus may, or may not, be congruent to the recurrent patterns of behavior, which can be observed by researchers. This proposed turn in dynamic capability research is inspired by recent work on organization design that extends Simon's (1996) pioneering ideas in this area (e.g., Romme and Endenburg 2006), acknowledging a distinction between espoused beliefs and behavior and actual patterns of behavior that has not been acknowledged by evolutionary economics. As such, we extend the current literature on dynamic capability that largely draws on economics as its disciplinary base (Helfat et al. 2007), with theory on design. This approach may also facilitate practitioner-academic projects set up to make firms more effective in reconfiguring their operations and competencies.

4.2 The Nature of Dynamic Capability

4.2.1 Foundations of Dynamic Capability

Teece et al. (1997) introduced the concept of dynamic capability to explain a firm's competitive advantage. In their view, competitive advantage stems from high-performance routines operating inside the firm, shaped by distinctive organizational processes, asset positions and evolutionary paths (Teece et al. 1997). These high-performance routines constitute dynamic capability, defined as: "the firm's ability to integrate, build, and reconfigure internal and external competences to address rapidly changing environments" (Teece et al. 1997: 516). In their view, dynamic capabilities are detailed and have rather predictable outcomes. In changing environments competitive advantage can thus be build through reshaping existing (tangible and intangible) resources and capabilities, and through creating new ones (Teece et al. 1997; cf. Zollo and Winter 2002). Schreyögg and Kliesch-Eberl (2007) describe Teece et al.'s view as the integrated dynamization approach, where dynamic capabilities include adapting, integrating and reconfiguring integrated clusters of resources and capabilities to match the changing environment.

Another important view on dynamic capability is that of Eisenhardt and Martin (2000), who treat dynamic capabilities as capabilities that shape a firm's resource position (e.g., capabilities in firm acquisition, alliancing, product development, strategic decision making). Eisenhardt and Martin (2000: 1107) adopt the following definition of dynamic capability: "the firm's processes that use resources – specifically the processes to integrate, reconfigure, gain and release resources – to match and even create market change. Dynamic capabilities thus are the organizational and strategic routines by which firms achieve new resource configurations as markets emerge, collide, spit, evolve, and die." In Eisenhardt and Martin's (2000) view, dynamic capabilities are idiosyncratic and linked to market dynamism, exhibiting different features in two types of markets. In a moderately-dynamic market, dynamic capabilities resemble the traditional conception of capabilities as complicated, detailed and analytic. In high-velocity markets, however, dynamic capabilities tend to involve simple, experiential and unstable capabilities (Eisenhardt and Martin 2000). In addition, they argue that dynamic capabilities can be a source of competitive advantage if they are applied sooner, more straightforwardly, and more fortuitously than competition to create bundles of resources. As such, Eisenhardt and Martin (2000) suggest dynamic capabilities are specific capabilities that embrace not only detailed, analytic capabilities but also simple, experiential ones. Schreyögg and Kliesch-Eberl (2007) argue that Eisenhardt and Martin's (2000) view draws on a radicalization approach, in which radical dynamic capabilities serve to master high-velocity markets by linking and selecting capabilities to continuously create new combinations of resources.

4.2.2 What Dynamic Capability Is Not

A clear definition of dynamic capability also delineates what it is not. Several scholars have attempted to explain why dynamic capability is equivalent to, or differentiates from, problem solving routines. This discussion arises from Petroni (1998), who argues that routines adopted by firms in problem solving activities are the essence of dynamic capability. In 2003, Winter published a paper that introduces the concept of ad hoc problem solving, defined as non-routine and non-repetitious change activities, typically appearing as a response to relatively unpredictable events. According to Winter (2003), the ability to solve problems does not necessarily imply a dynamic capability (i.e., the ability to change the way the firm solves its problems). In fact, dynamic capabilities may be quite rare (Winter 2003). In this respect, Winter (2003) contrasts the cost structure of dynamic capabilities with that of ad hoc problem solving. Dynamic capabilities involve long-term commitments to specialized resources, for example, new product development or account management. The ability to sustain a particular approach and commitment to, for example, account management depends to some extent on continuity in staff experience, information systems, and client networks (cf. Winter 2003). However, Winter (2003) argues that the costs of ad hoc problem solving largely disappear if there is no problem to solve. These costs, if any, tend to be opportunity costs of staff with alternative productive roles in the organization (Winter 2003). The fundamentally different cost structures between dynamic capabilities and ad hoc problem solving may explain why dynamic capabilities tend to be rare and ad hoc problem solving tends to prevail in many firms.

Dynamic capabilities are also different from operating routines and processes (i.e., the firm's primary processes such as product development, production and sales); that is, dynamic capabilities govern the rate of change of operating routines and processes (Collis 1994). More specifically, Winter (2003) argues that dynamic capabilities operate to extend, modify or create ordinary (substantive) capabilities (i.e., operating routines and processes). In addition, Zollo and Winter (2002: 340) define dynamic capability as: "a learned and stable pattern of collective activity through which the organization systematically generates and modifies its operating routines in pursuit of improved effectiveness." In other words, we distinguish here between changes in operations such as rearranging chairs on a sinking ship (i.e., addressing symptoms) and those that change course and avoid hitting an iceberg (i.e., addressing causes). The definition of Zollo and Winter (2002) further implies dynamic capabilities consist of patterned organizational behavior that companies invoke on a repeated rather than idiosyncratic basis (cf. Cepeda and Vera 2007; Helfat et al. 2007). Although many scholars define dynamic capabilities as high-level routines (e.g., Teece et al. 1997), which implicates repetitive action patterns, only few definitions of dynamic capability *explicitly* incorporate the requirement of repetition (e.g., Wang and Ahmed 2007; Zollo and Winter 2002).

4.2.3 Dynamic Capability and the Tautology Problem

Dynamic capability involves three levels of analysis: its output or outcome, the capability itself, and what makes up the capability. The current literature tends to be unclear about the level of analysis. Moreover, the levels of analysis are interwoven. For example, dynamic capability and firm performance are intertwined in many studies. Because dynamic capabilities "pursue improved effectiveness" (Zollo and Winter 2002: 340), dynamic capability is considered to improve (1) financial performance in terms of return on assets and return on sales and/or (2) business performance in terms of market share, sales growth, diversification, and product development. This may raise a tautology problem in measuring dynamic capabilities, particularly when dynamic capability is inferred from successful firm performance: if the firm performs well, it apparently possesses dynamic capability; if performance is not superior, then the firm apparently scores low on dynamic capability (Zahra et al. 2006).

Eisenhardt and Martin (2000) first attempted to untangle dynamic capabilities from firm performance in order to solve the tautology problem. They state that dynamic capabilities are identifiable, specific processes that are neither vague nor tautological. Zott (2003) also makes the distinction between dynamic capabilities and firm performance, by synthesizing Zollo and Winter's (2002) work into an evolutionary model in which change processes (i.e., dynamic capabilities) operate on a firm's resource position, which then determines its performance in a competitive marketplace. Other scholars who focus on this issue are Zahra et al. (2006) and Helfat et al. (2007). They deliberately attempt to decouple the definition and measurement of dynamic capability from financial and business performance. For example, Zahra et al. (2006) describe dynamic capabilities as the abilities to reconfigure a firm's resources and routines in the manner its principal decision-maker(s) envisioned it. Similarly, Helfat et al. (2007) describe dynamic capability as the firm's capacity to purposefully create, extend, or modify its resource base, acknowledging that change in the resource base of an organization implies only that the organization is doing something different, but not necessarily better than before. Helfat et al. (2007) further propose the notion of evolutionary fitness, referring to how well a dynamic capability enables an organization to make a living by creating, extending, or modifying its resource base (i.e., living long term). By contrast, technical fitness refers to how effectively a capability performs its intended function (i.e., intended on the short term). Technical fitness has two dimensions: quality and cost, that is, how well the capability performs respectively how much it costs to perform at a certain level (Helfat et al. 2007). Evaluating technical fitness, however, needs to take place at an ad hoc basis, which could raise a new tautology problem.

A small number of *empirical* studies avoid firm performance indicators in measuring dynamic capability (Daniel and Wilson 2003; Newbert 2005; Verona and Ravasi 2003; Wilson and Daniel 2007). But the approach taken in most empirical studies does raise the tautology problem (e.g., Arthurs and Busenitz 2006; Iansiti and Clark 1994; Wilkens et al. 2004). For example, Arthurs and Busenitz (2006) measure dynamic capability in terms of the one year risk-adjusted stock price returns (as the performance indicator).

However, when firm performance is (conceptually) decoupled from dynamic capability, other issues remain. Defining dynamic capability in terms of for example specific behaviors rather than an ability to accomplish something, is as tautological as defining it in terms of firm performance. If a dynamic capability is an ability to do something (cf. Teece et al. 1997), then logically, it could change its meta-rules and still have the capability, even if the rules and behaviors change. An admittedly feeble analogy is a pitcher in baseball who has a really low Earned Run Average (ERA) (i.e., output or outcome of a capability). This could be based in a capability of throwing very fast balls, or in a capability to throw a variety of pitches and sequence them, so that hitters are taken by surprise (i.e., the capability itself). Those capabilities, in turn, are grounded in specific actions and behaviors (i.e., what makes up the capability).

An example taken from the field of individual learning and training illustrates how one might overcome such tautology problems. To demonstrate whether what was learned in a specific training is also actually applied on the job, two steps have to be taken. First, one needs to demonstrate behavioral changes in on-the-job behavior and second, one needs to demonstrate that such changes are due (at least partly) to the specific training the employee was exposed to (Cascio 1998). Applying this to dynamic capability, researchers need to demonstrate that changes in operating routines and processes are due to at least one dynamic capability, instead of solely examining if the existence of one or more dynamic capabilities leads to successful organizational firm performance, versus if underperformance results from the absence of such capabilities. In this respect, a fundamental challenge is to develop measures of dynamic capability that are grounded in existing theory, are empirically straightforward and valid (i.e., do not include direct or indirect measurements of firm performance, or specific rules and behaviors), and serve to help practitioners make their organizations more effective in revising and reconfiguring resources. Therefore, we propose a new definition of dynamic capability.

4.2.4 Defining Dynamic Capability

The previous review of what dynamic capability is (not) implies the following components of our definition of dynamic capability: dynamic capability involves those capabilities in a firm ...

- that convey *deliberate* knowledge (among its key agents), invoked on a repeated basis (e.g., Cepeda and Vera 2007; Helfat et al. 2007; Sher and Lee 2004; Zollo and Winter 2002), on how to question purpose and effectiveness of the resource base (e.g., Helfat et al. 2007; Winter 2003; Zollo and Winter 2002);
- that serve to generate and modify operating routines and processes (e.g., Eisenhardt and Martin 2000; Teece et al. 1997; Zollo and Winter 2002) to address changing environments and/or create market change (e.g., Eisenhardt and Martin 2000; Teece et al. 1997).

We thus define dynamic capability as capabilities *that convey deliberate knowledge, invoked on a repeated basis, on how to question purpose and effectiveness of the resource base; this deliberate knowledge serves to generate and modify operating routines and processes to address changing environments and/or create market change.* This implies the routinized ability to raise particular questions regarding the effectiveness of particular operating routines and processes can be regarded as a dynamic capability.

The proposed definition imposes boundaries on which capabilities can be understood as being dynamic in nature, which has obvious implications for what should not be interpreted as a dynamic capability. In particular, our definition, more explicitly than in some other studies (e.g., Eisenhardt and Martin 2000; Teece et al. 1997), implies that routinized capabilities with a low level of awareness are not understood as dynamic in nature. Similarly, a dynamic capability that subsequently matures and becomes more habitual and therefore requires less and less conscious thought (cf. Helfat and Peteraf 2003), as such breaks down as a dynamic capability (it may constitute a growing capability that is operational in nature). In addition, our definition implies that dynamic capability is likely to have a positive effect on firm performance. However, it does *not* imply that firm performance automatically increases as a result of developing a dynamic capability. In this respect, firm performance depends on other factors affecting performance, such as market and competitive conditions beyond the control of the firm.

4.3 Unpacking Dynamic Capability

Given our definition of dynamic capability, the question arises how scholars engaging in dynamic capability research can respond to practitioners who would like to build this kind of capability within their firm. As such, we need to unpack the notion of dynamic capability (cf. Teece 2007; Eisenhardt and Martin 2000). Teece (2007) adopted a micro-practice perspective enabling a deeper understanding of the structural, cognitive and behavioral constituents of dynamic capabilities. Eisenhardt and Martin (2000) also anticipate this kind of challenge by identifying dynamic capabilities as concrete business operations in the form of specific capabilities by which managers alter their resource base. In this respect, the existing literature focuses on dynamic capabilities as *recurrent patterns of behavior* in organizational and strategic processes over time, which refer to a firm's managerial and organizational processes (e.g., resource allocation) rather than the primary processes such as product development, production and sales (i.e., operating routines and processes) (e.g., Helfat et al. 2007; Rindova and Kotha 2001; Zollo and Winter 2002). In this chapter, we extend Eisenhardt and Martin's proposal by differentiating dynamic capabilities in recurrent patterns of behavior and so-called *design rules*.

The design perspective adopted here is inspired by recent work that extends Simon's (1996) pioneering ideas in this area (e.g., Brusoni and Prencipe 2006; Dunbar and Starbuck 2006; Romme and Endenburg 2006). Simon (1996) argues that creating new organizational designs or redeveloping existing ones asks for principles from which the rest will develop (Simon 1996). Romme and Endenburg (2006) describe design rules as any coherent set of guidelines for designing and developing organizations, grounded in a related set of construction principles (i.e., any coherent set of normative ideas and propositions for producing new organizational structures and forms and redeveloping existing ones). Design rules serve as a heuristic device that describes ideas and intentions underlying a particular organizational design and helps to make sense of the processes produced by this design and any changes the design needs to undergo (Romme and Endenburg 2006).

Romme (2003) and Van Aken (2004) suggest the notion of design rules serves to connect the largely descriptive and explanatory nature of academic (management) research to the largely normative and pragmatic ways of reasoning and acting by practitioners. Their work emphasizes the importance of a certain level of awareness (e.g., among key agents, such as executives and middle managers) of the design rules because these rules represent (emerging or established) mental models of managers, engineers, sales people, and other agents (i.e., what people believe they should do). However, evolutionary economists (e.g., Heiner 1983) define rules more broadly, as depicting the repertoire of actions the agent actually engages in (but is not necessarily aware of). More particularly, agents develop or adopt rules when certain actions are unreliable in responding to the environment and they are therefore better off limiting or reducing their repertoire of actions (Heiner 1983; Langlois 1986). This may seem paradoxical, because it implies that both stable and volatile environments imply rule-following behavior. The behavioral dynamics, however, are different. A stable environment implies rather rigid and predictable behavior as an adaptation to the lack of environmental change. But once the environment becomes volatile, its demands begin to exceed the agent's ability to respond. This causes the agent to retreat to more predictable patterns of action (Heiner 1983; Langlois 1986). Our distinction incorporates both the awareness dimension (i.e., design rules) and the evolutionary dimension (i.e., recurrent patterns of behavior).

The complex relation between design rules driving dynamic capability and recurrent patterns of behavior exhibiting dynamic capability has two major dimensions: intentionality and legitimacy. The intentionality dimension is acknowledged by Augier and Teece (2007), who argue that dynamic capabilities include an organization's (non-imitable) ability to sense changing customer needs, technological opportunities, and competitive developments, as well as its ability to adapt to and possible shape, the business environment in a timely and efficient manner: "A significant element of intentionality is involved" (Augier and Teece 2007: 179). As such, we extend this perspective by arguing that the emerged or established mental models of managers, engineers, sales people and other agents (i.e., design rules) drive the (re)creation of dynamic capability as an artifact and sustain its viability over time. This is the intentionality dimension of the relation between design rules and recurrent patterns of behavior.

The legitimacy dimension refers to the support from external stakeholders (e.g., shareholders, suppliers, local community) as well as internal stakeholders (e.g., board of directors, union representatives, middle management, employees) for any dynamic capability. Achieving legitimacy is critical (Zahra et al. 2006), even if dynamic capability is underdeveloped. It is useful to draw an analogy with what Argyris, Putnam, McLain Smith (1985) call theory-in-use and espoused theory. Theory-in-use involves what people actually do (i.e., recurrent patterns of behavior) and espoused theory is what they say they do (i.e., design rules). There is ample evidence that people in organizational settings behave consistently with their mental models, but often do not act congruently with what they espouse (Argyris et al. 1985). In other words, what is espoused may, or may not, be congruent to the actual patterns of capabilities (e.g., observed by researchers).

Figure 4.1 summarizes our framework. In this framework, the main object of dynamic capability is the incumbent firm's operating routines and processes (Adner and Helfat 2003; Eisenhardt and Martin 2000; Wang and Ahmed 2007), implying that a dynamic capability serves to create new operating routines and processes or change existing ones. Moreover, the model in Fig. 4.1 implies that market and competitive conditions influence the emergence and evolution of dynamic capability (e.g., as input into discussions on business strategy), whereas the latter has an indirect impact on the market and competitive conditions by way of changes in the operating routines and processes. The evolutionary fitness between the operating routines and the market and competitive conditions over time determines firm performance (cf. Helfat et al. 2007). As such, the relationship between dynamic capability and long-term performance is an indirect one (Adner and Helfat 2003; Eisenhardt and Martin 2000; Wang and Ahmed 2007). In the remainder of this section we use this framework to develop a set of propositions for further research.

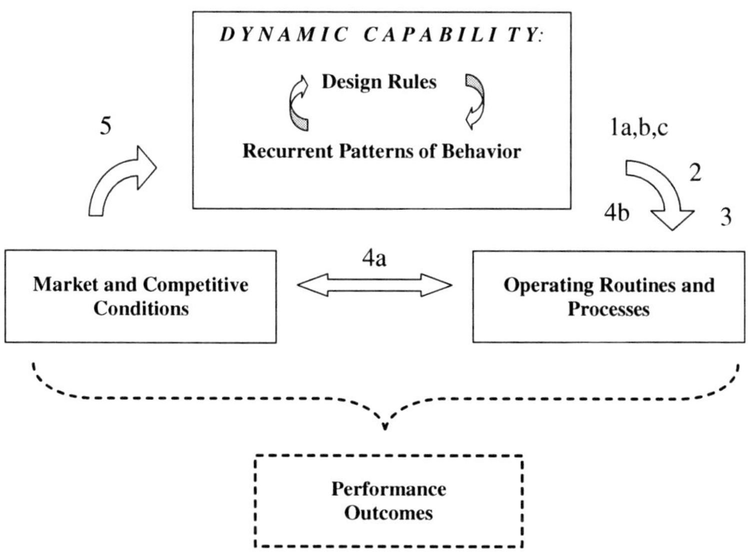

Fig. 4.1 Unpacking dynamic capability (numbers refer to propositions)

4.3.1 *Design Rules: Deliberate Knowledge on Questioning Purpose/Effectiveness*

The definition of dynamic capability we developed previously implies that a firm's performance in timely revising and reconfiguring its resource base, depends on deliberate knowledge (in terms of explicit design rules) and frequency (i.e., repetition) of invoking this knowledge (Helfat et al. 2007; Sher and Lee 2004; Zollo and Winter 2002). For example, Rubbermaid failed to adapt its "product innovation" strategy to the changing market conditions in the 1990s, as a result of the new CEO who came on board in 1992 and appeared to be "a desensitized leader who consistently missed the most telling signs of change in the industry and allowed his organization to become slow, unresponsive, and stagnant" (Helfat et al. 2007: 52). This episode in Rubbermaid's history illustrates how executives who miss any knowledge as well as experience in questioning the firm's resource base undermines the firm's capability to timely and effectively anticipate and respond to changes in market and competitive conditions (cf. Helfat et al. 2007). We therefore formulate the following claims:

Proposition 1a: The more deliberate knowledge the key agents in the incumbent firm have (in terms of explicit design rules) on how to question purpose and effectiveness of its resource base (D), the more effective they will be (i.e., in terms of setting right targets and choosing appropriate actions to achieve an overall goal) in changing its operating routines and processes (E).

Proposition 1b: The higher the frequency of invoking and applying this knowledge (F), the more effective the firm will be in (repeatedly) changing its operating routines and processes (E).

Proposition 1c: The causal claims in propositions 1a and 1b reinforce each other. That is, $E = f(D \times F)$, assuming all three variables are measured on a scale of 0 to 1.

Proposition 1c implies that at very high levels of both deliberate knowledge and the frequency of applying this knowledge, there are decreasing marginal returns in effectiveness. Another implication is that if either the deliberate knowledge or the frequency of its application is at a very low level (i.e., close to 0), increases in the level of the other variable will hardly affect effectiveness.

In the remainder of this section, we will refer to deliberate knowledge of questioning purpose and effectiveness of the firm's operating routines and processes in terms of design rules.

4.3.2 *Congruence of Design Rules and Recurrent Patterns of Behavior*

Adner and Helfat (2003) introduced the notion of dynamic managerial capabilities involving managerial cognitions, beliefs and mental models (i.e., design rules). These capabilities influence the strategic and operational decisions of managers (i.e., the recurrent patterns of behavior). We further elaborate this idea, by drawing on the notion of congruence between what managers believe and say they do and what they actually do.

A substantial incongruence, or gap, between espoused theory and actual behavioral patterns tends to arise when managers are severely challenged and engage in defensive reasoning. The incongruence between what is espoused and what is actually done is a major source of organizational inertia, because it inhibits both individual and organizational learning (Argyris et al. 1985). This phenomenon has been observed in many empirical studies of unsuccessful organizational and strategic change projects (e.g., Foss 2003; Helfat et al. 2007; Labianca et al. 2000). For example, the Danish firm Oticon (now William Demant Holding) tried to delegate decision-making to improve entrepreneurial capabilities and motivation at local organizational levels (Foss 2003). As such, Oticon's senior management deliberately attempted to build a credible commitment *not* to intervene in delegated decision-making (cf. the espoused design rules in this case). Nevertheless, frequent managerial meddling with delegated rights led to a severe loss of motivation (cf. the recurrent patterns of behavior) and caused Oticon to return to a much more conventional type of organization (Foss 2003).

By contrast, other studies suggest that organizational and strategic change occurs much more effectively if managers demonstrate a high level of congruence between promoting and practising the continuous search for new ideas and methods, trial and error experimentation, and so forth (e.g., Marcus and Anderson 2006; Rindova and Kotha 2001). For example, Rindova and Kotha (2001) observe that Yahoo! engages in continuous morphing of its form and function in the market, to the degree that it engages in self-organizing through reliance on simple organizational principles. An important difference with other cases (e.g., Oticon) appears to be the congruence between the principles espoused by the top management team and the behavior this team actually exhibits. Similarly, in a study of large firms in a variety of manufacturing industries Menguc and Auh (2006) observe that market orientation, as an espoused intention, does not qualify as a dynamic capability in itself. Market orientation only positively affects firm performance when it is complemented by practising continuous innovation and experimentation (Menguc and Auh 2006). In sum, this leads to the following proposition:

> Proposition 2: The greater the congruence between design rules for and the recurrent patterns of behavior in changing operating routines and processes, the more effectively the incumbent firm can engage in continuous experimentation and innovation and the less likely it will suffer from organizational inertia.

4.3.3 Codifying Distributed Learning and Control

In the context of knowledge codification (e.g., written documents for the purpose of storing knowledge), information technology is generally understood as an important device for enhancing dynamic capability (e.g., Sher and Lee 2004). There is an emerging body of evidence, however, that suggests knowledge codification may enhance operating routines and processes, but tends to undermine organizational and strategic transformation processes (e.g., Bhatt and Grover 2005; D'Adderio 2004; Mosey 2005).

An interesting example is the attempt to design and implement knowledge management practices at Infosys Technologies, the global software services firm. In the early 1990s Infosys started to transform employees' knowledge into an organization-wide resource that would "make every instance of learning within Infosys available to every employee" (Garud and Kumaraswamy 2005: 22). This project involved, amongst others, a central knowledge portal supported by e-mail, bulletin boards, and repositories for marketing, technical, and project-related information. To motivate employees to contribute content to the knowledge portal, Infosys implemented an incentive system involving monetary rewards or prizes, which produced several unintended consequences, such as information overload, decreasing quality and relevance of contributions, and a breakdown of the culture of freely sharing knowledge (Garud and Kumaraswamy 2005). The incentive system was therefore substantially modified and reduced, to decrease its negative impact on the informal culture of collaborative learning and knowledge sharing (Garud and Kumaraswamy 2005).

D'Adderio (2004) analyzed the influence of software on the dynamic capability of IT and other high-tech firms. She observes how the authority to change in these firms becomes distributed among engineers, software systems and design practices. D'Adderio (2004) finds that the disciplining action of software can be beneficial to certain, clearly structured processes for which requirements for stability and control are high; but it can represent a source of rigidity to other, more unstructured, functions and activities. An excessive emphasis on control may prevent the exploration of alternative technology configurations, as well as weaken the ability to incorporate heterogeneous knowledge inputs in the design (including inputs from customers and suppliers, and design feedback from other organizational functions and disciplines) (D'Adderio 2004).

In sum, this suggests that standardization and codification may be quite effective if applied to well-structured and unambiguous processes (e.g., a sequence of operations in a particular work setting), but less so for the more complex and ambiguous processes of distributed learning and control. This implies the following proposition:

> Proposition 3: Firms adopting design rules (for organizational and strategic change) that involve distributed learning and control will be more effective if the latter processes are not, or to a very limited extent, standardized and codified in software systems.

4.3.4 Market/Competitive Conditions and the Nature of Design Rules

As we have observed earlier, the behavioral dynamics of creating dynamic capability is fundamentally different in stable versus dynamic (market) conditions. Drawing on game theory, Langlois (1986) argues that increased flexibility is always desirable as the volatility of the environment increases. This increased flexibility tends to come in the form of more general actions, rather than specialized actions, that can be applied across a variety of environmental states (Langlois 1986).

This theoretical finding corresponds with the observation by Eisenhardt and Martin (2000) that routines in very volatile markets are purposefully simple, although not completely unstructured. Langlois' (1986) proposition also resonates with Rindova and Kotha (2001), who observe that the dynamic capability of Yahoo! is emergent and evolving, and grounded in open-ended organizing principles (cf. design rules). For example, the vice president of business development of Yahoo! is reported to describe the principles for creating partnerships as follows: "Put the product first. Do a deal only if it enhances the customer experience. And enter no joint ventures that limits Yahoo!'s evolvability" (Girotto and Rivkin 2000; cited in: Rindova and Kotha 2001: 1274). Other design rules for dynamic capability at Yahoo! involve a decentralized structure emphasizing the autonomous action of individuals. These general, open-ended design rules appear to be at the core of Yahoo's dynamic capability as a simple set of rules that are repetitively applied to changing operating routines and processes (Rindova and Kotha 2001).

Another example of a deliberate attempt to develop design rules for dynamic capability is Romme and Endenburg's (2006) study of circular organizing in more than thirty small and medium-sized firms, all situated in increasingly volatile environments. These organizations share a body of principles for increasing organizational capacity for self-regulation and learning. These principles involve a set of general heuristics regarding search, learning and hierarchy (e.g., "mistakes must be made") as well as a set of simple rules for organizational decision-making, governance and learning (e.g., "decisions on policy issues are taken by informed consent") (Romme and Endenburg 2006: 295–296). These design rules serve to create and sustain a particular type of communication and decision processes that leave the content of decisions and actions (to be) taken open and responsive (Romme and Endenburg 2006).

These findings suggest that open-ended design rules for dynamic capability (e.g., with regard to decentralized processes and local autonomy) apply to a broad set of environmental cues. By contrast, more closed design rules (e.g., "when acquiring another firm, first assess the target firm by means of our M&A protocol") apply to a specific category of environmental cues. In sum, we suggest that increased volatility in market and competitive conditions implies that open-ended design rules can be used, whereas increased stability has the opposite effect:

Proposition 4a: The more volatile the market and competitive conditions are, the more likely it is that changes in operating routines and processes will occur.

Proposition 4b: The latter relationship is reinforced by open-ended design rules for organizational and strategic change and undermined by relatively closed rules. That is, open-ended design rules for change are more effective in changing operating routines and processes than closed design rules (in volatile market and competitive conditions).

Zollo and Winter (2002) argue that the level of investment in developing dynamic capabilities will be the lowest when the firm draws on experience accumulation, as the learning then happens in an essentially semi-automatic fashion (e.g., learning-by-doing). This is therefore likely to be a valid approach in less volatile market environments. In more volatile environments, the learning investment is likely to be higher, particularly when the organization (or the relevant unit) relies on knowledge articulation (e.g., in meetings) to attempt to master or improve a certain activity because the organization will have to incur costs due to the time and energy required for people to meet and discuss their respective experiences and beliefs. This implies the necessity of a high level of cognitive effort because there is a certain level of understanding of the causal mechanisms intervening between the actions required to execute a certain task and the performance outcomes produced. According to Zollo and Winter (2002), such articulation efforts can produce an improved understanding of the new and changing action-performance links, and as such result in adaptive adjustments to the existing sets of routines or in enhanced recognition of the need for more fundamental change. Moreover, the learning investment and cognitive effort will be the highest for knowledge codification because when executing the task, people not only have to meet and discuss, but they also have to actually develop a document or a tool (e.g., manual or piece of software) aimed at distilling the insights achieved during discussions. If such a document or tool already exists, one has to decide whether and how to update it, and then to actually do the update (Zollo and Winter 2002). Mosey's (2005) study of five small- and medium-sized firms engaging in new-to-market product development illustrates this. In firms effectively engaging in product development, managers empower cross-functional teams to evaluate new technologies with a substantial number of external partners, but also systematize and codify learning between projects (Mosey 2005). This leads to the following claim:

Proposition 5: The more knowledge intensive the operating routines and processes required by market conditions and industry standards are, the more effectively the incumbent firm can develop and apply design rules (for organizational and strategic change) that draw on knowledge articulation and codification.

These five propositions constitute a preliminary set of causal claims. In this respect, senior management faces the need to develop knowledge on how to question purpose and effectiveness, be congruent in terms of talk and walk, facilitate distributed learning and control without overreliance on standardization and codification, create a balance between the level of environmental volatility and the open-ended nature of design rules, and finally, align the knowledge intensity of the firm's operations with the level of articulation and codification of organizational and strategic change projects. The propositions 1 and 2 are rather novel, while propositions 3, 4 and 5 serve to summarize what previous studies imply. Some of these generative processes can be inherently antagonistic (such as those in propositions 3 and 5), which makes building and sustaining dynamic capability indeed a major challenge.

4.4 Discussion and Conclusion

4.4.1 Informing the Theory and Practice of Organizational Design

We have advocated a theoretical perspective in this chapter that ties previous research together as well as extending it. The proposed definition of dynamic capability that is grounded in a review of the literature, provides a starting point for scholars who wish to operationalize the notion of dynamic capability without producing the tautology problem. In addition, the design-based theory described in the previous sections allows for a variety of design rules and related operating routines and processes that scholars and practitioners can draw on, by focusing on the interaction and co-evolution between dynamic capability, operating routines and processes, and market and competitive conditions (cf. Fig. 4.1). As such, the approach proposed may facilitate practitioner-academic projects, particularly those intended to make firms more effective in responding to and anticipating changes in external conditions as well as reconfiguring their operating routines and processes.

The design-based approach we have advocated implies a fundamental extension of the evolutionary argument that currently prevails in the dynamic capability literature. In particular, collaborative research with senior executives that uncovers and codifies the espoused and behavioral dimensions of dynamic capability may serve to enhance this capability (cf. Rindova and Kotha 2001). The theory developed in this chapter explicitly acknowledges the possible gap between intended rules and observed patterns of behavior. Evidently, the sensitivity of the issues addressed here may create difficulties in obtaining access to empirical sites. In this respect, scholars will be more successful if they develop long-term partnerships with practitioners and their organizations and are able to deliver tangible results in terms of codified (future) practices.

4.4.2 Suggestions for Future Research

Future work should draw on further elaborating and explaining the relation between dynamic capability and external influences. In particular, we need to study the generative forces that drive the co-evolution between operating routines and processes and market and competitive conditions, and its effect on performance outcomes (cf. Fig. 4.1). The notions of technical and evolutionary fitness, suggested by Helfat et al. (2007), provide an appropriate framework for future work in this area. In addition, studies that draw on both quantitative and qualitative data will be more likely to advance the theory of dynamic capabilities. Given the lack of consensus on key concepts and measurements, scholars cannot yet exclusively rely on quantitative data to establish causal relationships between dynamic capabilities, operating routines and processes, market and competitive conditions, and performance outcomes. Future research may thus benefit from adopting other research methods as well, such as simulation modeling (e.g., Lenox 2002). Although biases and limitations of the modeler might be simulated as well, an advantage of simulation modeling is that experimenting in a model environment can provide valuable insights into the complex feedback loops linking the antecedents, processes and outcomes of dynamic capability. In addition, simulation modeling may capture the evolution of dynamic capabilities more effectively than any other research method.

4.4.3 Conclusion

Building and sustaining dynamic capability provides a significant challenge, for practitioners trying to create such a capability as well as for researchers attempting to understand the process of capability formation. In this chapter we developed an integrative framework of this phenomenon. This framework suggests how dynamic capability can be unpacked and empirically studied. As such, it provides a starting point for future theoretical work as well as empirical tests that may advance knowledge on dynamic capabilities as key drivers of long-term business performance.

References

Adner R., Helfat CE (2003) Corporate effects and dynamic managerial capabilities. Strategic Management Journal 24: 1011–1025.

Aken JE van (2004) Management research based on the paradigm of the design sciences: The quest for field-tested and grounded technological rules. The Journal of Management Studies 41 (2): 219–246.

Argyris C, Putnam R, McLain Smith D (1985) Action Science: Concepts, methods, and skills for research and intervention. San Francisco: Jossey-Bass.

Arthurs JD, Busenitz LW (2006) Dynamic capabilities and venture performance: The effects of venture capitalists. Journal of Business Venturing 21: 195–215.

Augier M, Teece DJ (2007) Dynamic capabilities and multinational enterprise: Penrosean insights and omissions. Management International Review 47 (2): 175–192.

Bhatt GD, Grover V (2005) Types of information technology capabilities and their role in competitive advantage: An empirical study. Journal of Management Information Systems 22: 253–277.

Brusoni S, Prencipe A (2006) Making design rules: A multidomain perspective. Organization Science 17 (2): 179–189.

Cascio WF (1998) Applied psychology in human resource management, 5th edn. New Jersey: Prentice Hall.

Cepeda G, Vera D (2007) Dynamic capabilities and operational capabilities: A knowledge management perspective. Journal of Business Research 60 (5): 426–437.

Collis DJ (1994) Research note: How valuable are organizational capabilities? Strategic Management Journal 15: 143–152.

D'Adderio L (2004) Inside the virtual product: How organizations create knowledge through software. Cheltenham: Edward Elgar.
Daniel EM, Wilson HN (2003) The role of dynamic capabilities in e-business transformation. European Journal of Information Systems 12: 282–296.
Dunbar RLM, Starbuck WH (2006) Learning to design organizations and learning from designing them. Organization Science 17 (2): 171–178.
Eisenhardt KM, Martin JA (2000) Dynamic capabilities: What are they? Strategic Management Journal 21: 1105–1121.
Foss NJ (2003) Selective intervention and internal hybrids: Interpreting and learning from the rise and decline of the Oticon spaghetti organization. Organization Science 14: 331–349.
Garud R, Kumaraswamy A (2005) Vicious and virtuous circles in the management of knowledge: The case of Infosys Technologies. MIS Quarterly 29 (1): 9–33.
Girotto J, Rivkin J (2000) Yahoo! Business on internet time (Harvard Business School Case #9-7000-013). Boston: Harvard Business School Press.
Heiner RA (1983) The origin of predictable behavior. The American Economic Review 73 (4): 560–595.
Helfat CE, Finkelstein S, Mitchell W, Peteraf MA, Singh H, Teece DJ, Winter SG (2007) Dynamic capabilities: Understanding strategic change in organizations. Malden: Blackwell.
Helfat CE, Peteraf MA (2003) The dynamic resource-based view: Capability lifecyles. Strategic Management Journal 24: 997–1010.
Henderson RM, Clark KB (1990) Architectural innovation: The reconfiguration of existing product technologies and the failure of established firms. Administrative Science Quarterly 35: 9–30.
Iansiti M, Clark KB (1994) Integration and dynamic capability: Evidence from product development in automobiles and mainframe computers. Industrial and Corporate Change 3 (3): 557–605.
Javidan M (1998) Core competence: What does it mean in practice? Long Range Planning 31 (1): 60–71.
Kogut B, Zander U (1992) Knowledge of the firm, combinative capabilities, and the replication of technology. Organization Science 3: 383–397.
Labianca G, Gray B, Brass DJ (2000) A grounded model of organizational schema change during empowerment. Organization Science 11: 235–257.
Langlois RN (1986) Coherence and flexibility: Social institutions in a world of radical uncertainty. In: Kirzner I (eds), Subjectivism, intelligibility, and economic understanding: Essays in honor of the eightieth birthday of Ludwig Lachmann. New York: New York University Press.
Lenox M (2002) Organizational design, information transfer, and the acquisition of rent-producing resources. Computational & Mathematical Organization Theory 8:113–131.
Marcus AA, Anderson MH (2006) A general dynamic capability: Does it propagate business and social competencies in the retail food industry? Journal of Management Studies 43 (1): 19–46.
Menguc B, Auh S (2006) Creating a firm-level dynamic capability through capitalizing on market orientation and innovativeness. Journal of the Academy of Marketing Science 34 (1): 63–73.
Mosey S (2005) Understanding new-to-market product development in SMEs. International Journal of Operations & Production Management 25 (2): 114–130.
Newbert SL (2005) New firm formation: A dynamic capability perspective. Journal of Small Business Management 43 (1): 55–77.
Petroni A (1998) The analysis of dynamic capabilities in a competence-oriented organization. Technovation 18 (3): 179–189.
Priem R, Butler J (2001) Is the resource-based 'view' a useful perspective for strategic management research? Academy of Management Review 26: 22–40.
Prieto IM, Easterby-Smith M (2006) Dynamic capabilities and the role of organizational knowledge: An exploration. European Journal of Information Systems 15: 500–510.
Rindova VP, Kotha S (2001) Continuous "morphing": Competing through dynamic capabilities, form, and function. Academy of Management Journal 44 (6): 1263–1280.

Romme AGL (2003) Making a difference: Organization as design. Organization Science 14 (5): 558–573.
Romme AGL, Endenburg G (2006) Construction principles and design rules in the case of circular design. Organization Science 17 (2): 287–297.
Schreyögg G, Kliesch-Eberl M (2007) How dynamic can organizational capabilities be? Towards a dual-process model of capability dynamization. Strategic Management Journal 28: 913–933.
Sher PL, Lee VC (2004) Information technology as a facilitator for enhancing dynamic capabilities through knowledge management. Information & Management 41: 933–945.
Simon HA (1996) The sciences of the artificial, 3rd edn. Cambridge: MIT Press.
Sirmon DG, Hitt MA, Ireland RD (2007) Managing firm resources in dynamic environments to create value: Looking inside the black box. Academy of Management Review 32 (1): 273–292.
Teece DJ (2007) Explicating dynamic capabilities: The nature and microfoundations of (sustainable) enterprise performance. Strategic Management Journal 28: 1319–1350.
Teece DJ, Pisano G, Shuen A (1997) Dynamic capabilities and strategic management. Strategic Management Journal 18 (7): 509–533.
Verona G, Ravasi D (2003) Unbundling dynamic capabilities: An exploratory study of continuous product innovation. Industrial and Corporate Change 12 (3): 577–606.
Wang CL, Ahmed PK (2007) Dynamic capabilities: A review and research agenda. International Journal of Management Reviews 9 (1): 31–51.
Wilkens U, Menzel D, Pawlowsky P (2004) Inside the black-box: Analysing the generation of core competencies and dynamic capabilities by exploring collective minds. An organisational learning perspective. Management Review 15 (1): 8–26.
Wilson H, Daniel E (2007) The multi-channel challenge: A dynamic capability approach. Industrial Marketing Management 36: 10–20.
Winter SG (2003) Understanding dynamic capabilities. Strategic Management Journal 24: 991–995.
Zahra SA, Sapienza HJ, Davidsson P (2006) Entrepreneurship and dynamic capabilities: A review, model and research agenda. Journal of Management Studies 43 (4): 917–955.
Zollo M, Winter SG (2002) Deliberate learning and the evolution of dynamic capabilities. Organization Science 13 (3): 339–351.
Zott C (2003) Dynamic capabilities and the emergence of intra-industry differential firm performance: Insights from a simulation study. Strategic Management Journal 24: 97–125.

Chapter 5
Predicting Organizational Reconfiguration

Timothy N. Carroll and Samina Karim

Abstract This chapter addresses the issue of structural change within for-profit organizations, both as adaptation to changing markets and as purposeful experimentation to search for new opportunities, and builds upon the "reconfiguration" construct. In the areas of strategy, evolutionary economics, and organization theory, there are conflicting theories that either predict structural change or discuss obstacles to change. Our aim is to highlight relevant theoretical rationales for why and when organizations would, or would not, be expected to undertake structural reconfiguration. We conclude with remarks on how these literatures, together, inform our understanding of reconfiguration and organization design and provide insights for practitioners.

Keywords Reconfiguration · Organization design · Structure · Restructuring · Reorganization · Organizational change · Configuration · Strategic choice

5.1 Introduction

In the fields of strategy and organization theory there is a large body of literature on organizational change. Research questions include what causes change, what deters or slows change, what constitutes change, and what are the consequences of change. In this chapter we focus on one element of change in particular – structural change within for-profit organizations. By changing the structural boundaries of the firm and further rearranging what resides within the firm, organizations may be able to realign themselves for better efficiency and effectiveness.

The study of changing organizational structure has received significant attention, stemming from Chandler's (1962) seminal work on the M-form organization. Chandler developed the strategy-structure paradigm, which proposed that firms should adapt their structures to best fit their strategic goals. This notion that changes

T.N. Carroll (✉)
Moore School of Business, University of South Carolina, Columbia, SC 29208, USA
e-mail: carroll@moore.sc.edu

A. Bøllingtoft et al. (eds.), *New Approaches to Organization Design*, Information and Organization Design Series 8, DOI 10.1007/978-1-4419-0627-4_5,

in strategy drive structural change has become a foundational tenet in organization studies. Various organization theorists in the contingency theory tradition have extended this argument of fit between the firm and its environment by also stressing the importance of fit within core organizational elements (for a review of the evolution of contingency theory see Donaldson 1990; for an integrated contingency framework see Burton and Obel 2004). There are few empirical studies (Bergh 1998; Keats and Hitt 1988), however, that find generalized support for environmental change causing structural change, and few papers simultaneously address both of these concepts – external environmental elements along with internal core elements.

Our purpose in this chapter is to review both the causes and obstructions to "structural reconfiguration" and to present relevant theories that may help us to make predictions about structural change. We begin by clarifying what we mean by structural reconfiguration. Next, we identify the theories and factors that predict reconfiguration, followed by those that hinder or do not predict it. Finally, we will integrate these perspectives and analyze overlap and gaps and conclude with discussion of directions for future research.

5.1.1 Defining Structural Reconfiguration

There are several terms that have been used (sometimes interchangeably and sometimes not) in the literatures to refer to structural change. Relevant terms include reorganization, restructuring, patching, and reconfiguration (Karim 2006).

"Reorganization" is a general term that can be applied to many contexts including those of individuals, resources, processes, tasks, or structures. Similarly, scholars have identified various forms of restructuring including financial, portfolio and organizational (Johnson 1996). The early strategy literature on diversification and portfolio management referred to restructuring as the process of entering and exiting business markets (e.g., Porter 1987). The concept of organizational "restructuring"[1] is closely aligned with the dynamic capabilities view of "patching" which is defined as "the strategic process by which corporate executives routinely remap businesses to changing market opportunities. It can take the form of adding, splitting, transferring, exiting, or combining chunks of businesses" (Eisenhardt and Brown 1999: 73–74). Organizations patch to realign themselves with their changing environment. Compared to organizational restructuring, patching is more proactive than reactive, is done more frequently, and often the changes are more incremental and evolving (Eisenhardt and Brown 1999).

The concept of "reconfiguration" is a broader construct that refers to firms adding to their current stock (of resources, units, and business activities), removing from this stock, and recombining what is within this stock (Karim and Mitchell 2004).

[1] Eisenhardt and Brown (1999) broadly refer to "reorganization" in their comparison to patching, however, we refer more specifically to their "reorganization" construct as that of "organizational restructuring."

In this chapter, we follow the definition in Karim (2006) and consider structural reconfiguration to encompass "the addition of units to the firm, deletion of units from the firm, and recombination of units within the firm such that resources and activities are still retained by the organization" (p. 799). Karim builds upon the notion of "patching" but adds that reconfiguration includes not only adaptation to match changing markets but also "a purposeful experimentation and search for new opportunities" (Karim 2006: 801).

5.2 Causes of Structural Reconfiguration

Our goal is to better understand the causes and impediments of structural reconfiguration and the associated theories that pertain to it. As the study of purposeful experimentation with structure is relatively new, we include studies of restructuring and patching to provide insight into our research question.

There has been a significant amount written about restructuring, yet there have been few large-sample empirical works. We are motivated by Hirsch and de Soucey (2006) who point out, "the term organizational restructuring is used sparingly, and its determinants are conceptually undertheorized" (2006: 174). In his review of past literature, Johnson (1996) studied one particular type of restructuring – the refocusing or downscoping of the firm by selling off business units – and identified five antecedents of refocusing. The antecedents are (1) environment, (2) governance, (3) strategy, (4) performance, and (5) financial restructuring. Since we are concerned with more than downscoping or a constriction of the organization's activities we extend Johnson's set of antecedents and causes of structural reconfiguration to also include the impact of life cycles (Section 2.6) and changes in technology or innovation in a major product (Section 2.7).

5.2.1 Environmental Change

Structural contingency theory specifies that an efficient organization will have structures that best fit the prevailing contingencies (Burton and Obel 2004; Donaldson 1995). These contingencies include contextual factors such as the firm's strategy, size, task uncertainty, and technology. These organizational characteristics, in turn, are affected by the environment in which the firm operates. Thus, in order to be effective, an organization's structure must fit its contingencies and its operating environment. As the environment changes and causes subsequent changes in the contingencies, the firm, in turn, must adapt to the environment (Donaldson 1990).

As environmental uncertainty increases, and with it task uncertainty, an organic structure characterized by loosely defined roles, dispersed knowledge, and the spontaneous cooperation of teams of experts is more effective (Burns and Stalker 1961). Lawrence and Lorsch (1967) specified that these greater rates of environmental change promote a more organic structure within certain parts of the organization compared to others, and that the structural and cultural differences between the departments (i.e., differentiation) makes coordination difficult (i.e., integration).

In contingency theory, environmental change causes a change in the more proximate organizational contingencies, such as the task environment and technology. When these contingencies change, an organization previously in fit moves into misfit and suffers declining performance as a result (Donaldson 2001; Gresov 1989; Nadler and Tushman 1984; Naman and Slevin 1993). Even a small number of misfits or misalignments may significantly compromise performance (Burton et al. 2002). This creates pressure for change that impels the organization to adjust via structural adaptation to move back into fit (Donaldson 1987).

Thus, the contingency literature suggests that structural reconfiguration is most likely when either a change in the environment or the organizational contingencies (such as the firm's strategy) cause an organization to fall into misfit, although the period of misfit may be lengthy (Donaldson 1987).

A second theoretical approach, mainly from the dynamic capabilities literature, concentrates on organizational forms suited for high velocity environments that require frequent adjustments and speedy adaptation (Eisenhardt and Brown 1999; Eisenhardt and Martin 2000; Galunic and Eisenhardt 1996, 2001; Helfat and Eisenhardt 2004; Teece et al. 1997). In their study of patching organizations, Eisenhardt and Brown stress that organizations in dynamic markets benefit from being the right size, where optimal size is determined by the ability of business units to be "small enough for agility and large enough for efficiency" (1999: 74). This is possible when divisional managers have the mind-set to track business metrics, realign over time and are able to make proactive decisions about structural change (Eisenhardt and Brown 1999). These ideas are reiterated by Huber (2004) who describes the conditions that will be faced by firms in the future and suggests what they may do to address these conditions. Specifically, he predicts that future firms will experience greater market velocity (i.e., more events per unit of time), turbulence (i.e., abrupt occurrences of events), and instability (i.e., shorter periods between technological breakthroughs). Under these circumstances of environmental dynamism, Huber (2004) proposes that firms will need to make decisions more frequently and rapidly; this may be achieved by increasing the number of "decision units" and the efficiency of decision making processes. Integrating these works suggests that firms in dynamic environments would benefit from reconfiguring their structures such that there are many units of smaller size. This is also a trend observed recently by scholars (Zenger and Hesterly 1997).

In a third stream of research, namely restructuring literature, several papers have empirically examined the impact of environmental change on structural reconfiguration. To better understand the logic of what is driving restructuring, Bergh (1998) studied firms' acquisition behavior during times of environmental uncertainty. He proposed that if the goal is to achieve efficiency then firms will acquire related businesses and divest unrelated businesses. What he finds, however, is that firms acquire *both* types of businesses – related and unrelated. This, he infers, indicates that during times of uncertainty, the desire to spread risk is greater than the incentive to focus on efficiency. These findings support an earlier study by Keats and Hitt (1988) who examined the effects of environmental instability on organizations' levels of divisionalization and size. They found that instability in the industries within which the manufacturing firms operated resulted in less divisionalization and larger unit size; stated differently, they found that environmental instability led to firms preferring fewer but larger units. These findings reiterate that during times of uncertainty, firms may prefer the safety of larger units within which activities may be cross-subsidized as the slack from one business activity benefits another.

The findings of these empirical studies that firms reconfigure to fewer, larger units in times of uncertainty or instability contrast with the predictions noted earlier that predict change toward more, smaller units that are flexible and agile. We will integrate these differences in Section 4.4 of our chapter.

5.2.2 Changes in Firm Governance

The impact of firm governance on structural reconfiguration appears indirectly via the composition of the board of directors and the study of executive turnover.

While the evidence for which characteristics of board composition predict good or bad governance is mixed (Dalton et al. 1998), studies have shown that poor board governance may lead to poorly conceived strategies as well as flaws in executing those strategies. Studies focusing on board composition and leadership structure, for example, have found little evidence of a systematic relationship between board characteristics and firm financial performance (Dalton et al. 1998). However, studies have shown that poor board governance, whatever its source, leads to high product diversification, which in turn leads to weaknesses such as high relative debt and low relative R&D spending (Hoskisson et al. 1994). This lack of investment in the future ultimately leads to poor performance which likely necessitates reconfiguration (we expand upon the effect of performance on structural reconfiguration in Section 2.4).

Although there has been significant study of executive turnover in the context of mergers and acquisitions (Walsh 1988, 1989; Walsh and Ellwood 1991), there are few studies that address the reverse relationship of how executive turnover affects structural reconfiguration. In his paper about strategic choice, Child (1972) considers how decision makers' interpretations as well as formations of dominant coalitions (Cyert and March 1963) around strategies and goals affect the choice of organizational structure. Thus, as executives change and dominant coalitions and interpretations change, so will the desired and chosen structures leading to reconfiguration. These shifts in power distribution are part of what constitute an "organizational transformation" in the punctuated equilibrium model (Romanelli and Tushman 1994). Galunic and Eisenhardt also refer to the influence of power; they note that corporate executives' roles extend beyond performance oversight and often entail the "recrafting of corporate architecture" since they are the ones with the power to make such changes (2001: 1246). Thus, executive turnover likely results in structural reconfiguration.

5.2.3 Changes in Strategy

The idea that strategy influences structure dates back to Chandler (1962). Chandler argued that the creation of a multi-divisional, M-form structure in modern firms was a structural innovation from what used to be the functional organization. In his view, implementing strategy (such as the expansion of volume, geographical dispersion, vertical integration, and/or product diversification) determines which type of structure (such as a centralized, functionally departmentalized structure or multidivisional, decentralized structure) a firm employs. As a precedent to contingency theory, Chandler proposed that the correct fit between the environment and a firm's strategy-structure composition would result in economic efficiency; thus his view was that the environment influences the strategy-structure relationship, which in turn influences performance. From Chandler's perspective, one would expect that if a firm chooses to change its boundary and the organization of what resides within the firm, it will have to make changes in its structure to implement the new strategy.

Later work supported the view that strategies influence organizational structure through "the particular coordinative, technical, and control problems they create" (Miller 1988: 281) and that high performance can be achieved when there is an appropriate fit between the firm's strategy and its choice of organizational structure and control systems (Hill et al. 1992). The unidirectional causal flow outlined by Chandler, however, has been challenged by those who argue that strategy, structure, and environment all influence each other (Miller 1988) rather than proceeding in one path only.

Recent work in dynamic capabilities has explored how changes in market activities lead to structural change. In their studies of a large Fortune 500 firm, Galunic and Eisenhardt (1996, 2001) traced how market responsibilities, or "charters," were assigned to or removed from business divisions and sometimes the boundaries of these divisions themselves would change. Building upon these works, Karim (2006) studied how firms recombine their business divisions. She found that firms entering (or deepening their involvement in) markets by pursuing acquisition strategies usually recombined those units eventually with other units within the firm. Based on the longevity of these recombined units, she infers that firms recombined unit structures to reap more value from acquisitions. Similar to this goal of reaping value from firm assets, Helfat and Eisenhardt (2004) highlight that firms may also change their structures (including the activities and resources within) when exiting markets to achieve economies of scope.

Other scholars have associated the joint change of strategy with changes in structure. A strategic change was defined by Romanelli and Tushman (1994) as the entry into or exit from market segments; this was also referred to as "strategic shifts" by Nadler and Tushman (1997). Structural changes were defined as changes in form (e.g., functional to divisional) or degree of (de)centralization (Romanelli and Tushman 1994). These joint changes, along with changes in power distribution, constitute what Romanelli and Tushman label a "revolutionary transformation" in their punctuated equilibrium model.

Poorly conceived or implemented strategies may also impel structural reconfiguration. For example, Rumelt (1974) found that firms that follow a related diversification strategy (which he termed "controlled diversity") outperformed firms that pursued an unrelated diversification strategy, and overdiversification (too much product diversification relative to competitors in the industry) is likely to drive downscoping or diverstiture of organizational units (Hoskisson et al. 1994).

Performance (either good or bad) whether the result of strategy or other factors also may act as an antecedent to structural reconfiguration, which we consider in the next section.

5.2.4 Changes in Performance

Organizational scholars have addressed the issue of how changing performance, both positive and negative, may affect structural change. In a review of restructuring studies, Hoskisson and Johnson state that "...the research to date suggests that the primary spark to restructuring is poor performance, possibly due to high levels of diversification" (1992: 626). Nadler and Tushman also address that poor performance may require firms to "redirect their efforts and apply resources differently" (1997: 49). This application of resources usually refers to how they are coordinated and/or integrated within the boundary of the firm. Contingency theorists proposed that firms could improve performance through higher levels of differentiation and integration (Lawrence and Lorsch 1967). Also, the influence and importance of performance metrics in shaping structure is highlighted in Mintzberg's (1980) synthesis of organization design; in it he refers to the importance of establishing structure to best coordinate businesses in "divisional form" organizations. Thus, poor performance may initiate reconfiguration to alter coordination and integration of business units, their resources and business activities.

Alternatively, firms with positive performance changes – experiencing the potential for growth as they have greater slack resources available to them – may also pursue structural reconfiguration (Penrose 1959). Both Chandler (1962) and Nadler and Tushman (1997) stress the importance of moving from informal mechanisms (that may suffice for small, entrepreneurial firms) to formal structures that are more efficient at managing tasks as organizations experience growth. Child (1972) stressed that executives are faced with the task of structuring growing organizations; in specific, they need to decide structural outcomes such as the allocation of resources or whether to split the firm into smaller units. Building on transaction cost arguments, Zenger and Hesterly attribute the trend toward smaller firms and units (which they refer to as "disaggregation") as "motivated by the powerful performance incentives that accompany small size" (1997: 209). Thus, growth opportunities need not be associated necessarily with larger structures; instead, firms may choose to reconfigure into a greater number of modular units.

We should note that performance can make changes either more or less likely (see Section 3.3 for a discussion). In general, when performance is below an aspiration level change becomes more likely, and change is less likely when performance exceeds expectations (Greve 2003). In the context of this paper, we expect that negative performance likely leads to structural reconfiguration, and though positive performance and subsequent growth also may result in reconfiguration, we expect that it will be to a lesser extent.

5.2.5 Diversification and LBOs

Recall that reconfiguration involves not only adding and changing what is in the firm but also removing unwanted stock from the firm. One potential cause for this downsizing (i.e., restructuring) is leveraged buyouts (LBOs) (Phan and Hill 1995). In LBOs, there is often an assumption that the firm has been inefficient in its deployment of resources, especially in the pursuit of too much diversification. LBOs also typically involve the issuance of large levels of debt to finance the purchase. As a result, both the strategic intent of the new management team, as well as the cash flow requirements of the new debt, push the firm to sell off some organizational units and combine others in the search for greater levels of efficiency.

Jensen (1986, 1988) and others (Hoskisson and Turk 1990; Ravenscraft and Scherer 1987) argue that many firms diversify beyond an optimally efficient point. From the perspective of the new owners, this provides an opportunity to refocus the organization and generate greater free cash flows. In addition, since the purchase is often financed through high levels of debt, managers must generate high levels of cash to service debt payments, further supporting a more focused approach.

Phan and Hill (1995) point out that the new debt payments motivate managers to "slash unsound investment programs, reduce overhead, dispose of assets that are more valuable outside of the company, and restructure an organization to increase accountability and control" (p. 706). The cash generated by selling assets and cutting investments "can then be used to reduce debt to sustainable levels, creating a leaner, more efficient and more competitive organization" (p. 706).

As the firm refocuses on a strategic core, reconfiguration is likely to occur as some organizational units are sold off and others are combined in the search for greater efficiencies, higher productivity, and higher profitability. For example, there is empirical evidence that after management buyouts firms often downsize (Liebeskind et al. 1992) and focus on a core set of related businesses (Seth and Easterwood 1993).

5.2.6 Life Cycles

A different tradition in organizational studies views structural change as a function of a natural pattern of growth and maturity. As industries evolve and firms grow and change according to a deterministic pattern, reconfiguration is the natural and inevitable result. Life cycle theories of organization change posit that "a program, routine, rule, or code exists in nature, social institutions, or logic that determines the stages of development and governs progression through the stages" (Van de Ven and Poole 1995). Thus, an organization may not change in response to a change in the environment, but rather it evolves as it progresses through stages of a life cycle.

Quinn and Cameron (1983), in their review of nine models of organizational life cycles, developed an integrated model with four general stages. New organizations begin in the first stage, called "creativity and entrepreneurship," which is characterized by little structure, planning or coordination. Informal communication, processes, and structure become more pronounced in the second "collectivity" stage. In the third "control" stage, an emphasis on efficiency and predictability drives formalization of rules and a stable structure. The fourth stage is characterized by adaptation and renewal through an "elaboration of structure." This framework provides a more detailed rationale for Blau's finding that as organizations grow in size their structures become more differentiated and elaborated (Blau 1970).

Thus, structural reconfiguration is most likely to occur during the later stages of the life cycle as the organization deals with the natural limitations of the earlier organizational forms. For example, in Greiner's (1972) model (one of the nine reviewed by Quinn and Cameron 1983), organizations advance through a series of five stages by solving the major organizational problem of the previous stage. In this model reconfigurations typically occur in stage 4 ("growth through coordination") as a response to a lack of integration across organizational subunits that arises as a result of stage 3 ("growth through delegation").

Similar to industry life cycles, product life cycles may also affect changes in firm structure. If the firm offers only one product, then the firm may be expected to undergo structural reconfiguration as the product matures. In the early stages of product development the focus is on innovation and flexibility, and thus organic structures are optimal (Burns and Stalker 1961; Hage and Aiken 1969; Lawrence and Lorsch 1967). As the product matures the emphasis becomes less on innovation and more on efficiency and price competition, encouraging a switch to a more mechanistic structure.

Donaldson (1985) addresses the situation where different products are at different stages in the product life cycle. When the firm offers multiple products the key issues relate to scale and differentiation. As the firm grows in scale then the structure typically becomes one of divisions, based on product classes, geographies, customer segments, or some other logic. The greater the differentiation between the products, the greater the likelihood is that the firm will move from a functional structure to semi-autonomous divisions (Khandwalla 1977; Rumelt 1974).

Whether approaching the life cycle from the perspective of products, technologies, organizations, or industries, the logic is similar. Growth and development creates pressure on existing structures and encourages structural reconfigurations that better meet the emerging needs (Cafferata 1982; Child and Kieser 1981; Greiner 1972; Kimberley and Miles 1980) of the organization.

5.2.7 Change in Technology or Innovation in Major Product

In contrast to incremental or evolving change, scholars have also addressed how a change in the underlying technology of a firm's major product influences the likelihood of a change in structure. The key concept in this line of work is the "dominant design" (Abernathy and Utterback 1978; Tushman and Murmann 1998). A dominant design comes into being when a new product coalesces on a standard set of features and technologies after a period of exploration of design alternatives in prior product variants (Utterback 1994: 24). A dominant design sets a new product standard that consumers prefer. The basis of competition then shifts to firms seeking to produce products based on the dominant design at the lowest cost (Utterback and Suarez 1993). Dominant designs then evolve from a period of substantial (revolutionary) change and variation into a more stable period of incremental (evolutionary) change and variation (Abernathy and Utterback 1978). The evolution of these dominant designs has implications for organizations as well, since the "requisite organizational forms differ between eras of ferment and eras of incremental change" (Tushman and Murmann 1998: 30). Thus, structural reconfiguration is especially likely when the dominant design undergoes a period of revolutionary change and the existing, often mechanistic organizational form (Burns and Stalker 1961) gives way to a more organic form that is better suited to periods of rapid technological variation.

Along with the changes in the underlying technology of the firm's major product, scholars have also noted that the modularity of a firm's new major product may affect structural change. Modularity refers to the extent to which components are more or less integrated and interdependent. This concept is applicable at both a micro and macro level; at a micro level of analysis it may be applied to the study of product architecture (Henderson and Clark 1990). At a macro level, studies have highlighted how modular product systems can lead to modular organizational systems (Baldwin and Clark 2000; Garud and Kumaraswamy 1995; Lei et al. 1996; Sanchez and Mahoney 1996; Schilling and Steensma 2001). A firm's level of modularity will lessen as its business units gain "synergistic specificity" (i.e., low separability) (Schilling 2000). Chesbrough and Teece (1996) recommend that firms organize based on the level of separability/integration of their innovations. In particular, they suggest that autonomous innovations be supported by decentralized structures, whereas systemic innovations that are interdependent upon other innovations are better off managed by centralized, integrated structures. Overall, we propose that firms may structurally reconfigure to achieve the desired level of integration and coordination that best suits a particular design or innovation.

5.3 Impediments to Structural Reconfiguration

The idea that organizations may resist change, even in the face of good reasons to do so, has a long history in organizational studies. Merton (1957) and Crozier (1964), for example, discussed the tendency of bureaucracies to become rigid in the application of rules and regulations. This rigidity constrains the organization's ability to change despite good reasons for it, such as internal growth or transformations in the environment.

More recent studies have extended this idea in more detail, developing theories for why changes like structural reconfiguration would not be expected to occur. These theories include the concepts of structural inertia, perceptions and aspirations, and threat-rigidity. We consider each in turn.

5.3.1 Inertia

Much work in the area of organizational restructuring emphasizes the ineffectiveness of restructuring (Bowman and Singh 1993). Hannan and Freeman (1984) propose that structural inertia is a property of all organizational forms. Inertia arises as a result of selection pressures requiring that organizational structures be reproducible, in order to accommodate pressures for high accountability and high reliability. Accountability and reliability are accomplished when organizational goals and purposes are institutionalized and when activities are routinized. Thus, organizations that are stable will tend to have lower failure rates than those that undertake change such as structural reconfiguration.

According to this view, reliability depends on reducing the variance in performance, rather than simply increasing its mean level. This, in turn, makes change such as structural reconfiguration unreasonably risky, even as an attempt to adapt to a changing environment, and thus inertia is a necessary consequence for firm survival. Amburgey et al. (1993) extend this reasoning and argue that change is risky because it destroys some of the firm's existing practices and familiar routines, which causes a loss of competence.

While the original ecological theories emphasized the reasons why organizations were unlikely to change, later work in this tradition showed that under conditions of severe environmental change – punctuated, rapid, and substantial change that threatens organizations with extinction – organizations that do change improve performance (Haveman 1992), especially if the changes build on established routines and competencies. In addition, routines, although typically seen as a source of stability, may also be a source of change (Feldman and Pentland 2003).

Similarly, work focused on the impact of discontinuous technological changes recognizes the potential for the technological change to overcome organizational inertia. Tushman and Murmann (1998) postulate that "shifts between eras of ferment and eras of incremental change seem to require discontinuous organizational change. It may be that discontinuous technological changes that are isolated to peripheral subsystems and/or single functions can be initiated by revolutionary subunit change. In contrast, discontinuous technological change in a core subsystem will be initiated by system-wide organizational change" (p. 36).

5.3.2 Perceptions and Aspirations

The concept of inertia highlights that firms may have the tendency not to structurally change. Inert organizations resist change, especially in their core elements, one of which is their structure (Hannan and Freeman 1984).

Greve (2003), building on the work of March (1981), incorporates the effect of inertia in his work on the effect of aspiration levels on the likelihood of change. Greve finds that the likelihood of change differs depending on three different types of search processes – slack, organizational, and problemistic search.

Slack search is a result of extra time and resources that are used for experimentation, such as engineers engaged in product development. Institutionalized search is a function of the work done by organizational units devoted to search activities, such as the activities of the R&D, market research, and strategic planning departments. Finally, problemistic search occurs as a response to an organizational problem (Greve 2003: 54).

Again, we view structural reconfiguration as the addition to the organization's current stock (of resources, units, and business activities), deletion from this stock, and recombination of what is within this stock (Karim and Mitchell 2004), as well as the purposeful experimentation and search for new opportunities (Karim 2006). Thus search, both problemistic and opportunistic, appears to be the most relevant to reconfiguration activities (Ciborra 1996; Karim 2006). Slack search may also impact the likelihood of experimentation with structural reconfiguration, since there may be less reticence by managers to try something new with assets that are beyond what is essential for core business operations.

When performance exceeds an aspiration level we expect that structural reconfiguration is less likely due to problemistic search (as this is of low priority) but may still occur as a result of slack search.

When performance is below an aspiration level change becomes more likely, although the probability of change is held down to some extent by organizational inertia. In other words, absent the effect of inertia, the probability of change would become consistently less likely as performance improves at a constant rate or slope. Due to inertia the expected probability of change when performance is below expectations is still greater than when performance is above expectations, but the rate of increase in the likelihood of change is lower than would occur without inertia.

5.3.3 Threat-Rigidity

Greve also notes the possibility that low performance may be interpreted as a threat to the organization, and that threat-rigidity may occur (Greve 2003). The threat-rigidity hypothesis (Ocasio 1995; Staw and Ross 1987; Staw et al. 1981) suggests that "executives faced with threats perceive that they have little control over the situation and face the risk of a negative outcome" (Chattopadhyay et al. 2001: 939).

Threat-rigidity differs from regular performance feedback because the feared level of performance becomes more salient to the decision maker than the hoped-for aspiration level (Greve 2003; Lopes 1987; March and Shapira 1992). Lant and Hurley (1999) showed that threat-rigidity responses did indeed occur when performance was just below the aspiration level.

Threat-rigidity implies that organizations facing a situation such as environmental change may rely on standard responses, even when these responses are inappropriate (Staw et al. 1981). Two mechanisms may explain this phenomenon. First, an environmental change or threat may reduce information processing. Second, an environmental change may cause a constriction in control, where decision making becomes more centralized and concentrated. Thus, managerial perceptions and the processes by which information is collected and interpreted can have a large impact on organizational actions or lack of action (Hambrick and Mason 1984; Starbuck and Milliken 1988; Thomas et al. 1993).

5.4 Looking Forward

5.4.1 Informing the Theory of Organization Design

From our review of the literature, we find that there are several theories and factors that predict structural change and others that do not. Building off of Johnson's (1996) review, we categorize the main initiators of structural change as the environment, governance, strategy, performance, diversification and LBOs, firms' life cycles, technology and innovation. Literatures that have revealed these antecedents include contingency theory, organization design theory, dynamic capabilities perspective, strategy-structure paradigm, resource based view, and studies of diversification and mergers and acquisitions. Complementing these works are those that highlight the impediments to structural change; these include inertia, perceptions, aspirations, information processing, and path dependence. Supporting literatures include studies of population-ecology, evolutionary economics, and threat-rigidity theory.

The two factors contributing to structural change and receiving the most attention in academic work are the environment and strategy. Scholars examining the effect of strategy on structure are, for the most part, in agreement that as strategy changes firms are likely to change their structures to better implement the new strategy. Strikingly different is the case, however, for the environment's effect on structural change. Although several case-studies of firms find evidence of turbulent environments leading to patching and reconfiguration (Eisenhardt and Brown 1999; Galunic and Eisenhardt 1996, 2001; Karim and Mitchell 2004), the few empirical studies that exist question the generalizability of these cases and find evidence of firms being more risk averse during times of instability (Bergh 1998; Keats and Hitt 1988).

How can we make sense of these conflicting findings? We know that firms which change structure often need an infrastructure to support the process and their managers need to have a "distinct mind-set" (Eisenhardt and Brown 1999). Could it be that the executives of these firms are expected or encouraged to experiment with structure? Drawing on the findings of Keats and Hitt (1988) and Bergh (1998), is it that these firms that do reconfigure often are less risk averse? Or is it that they have reconfiguration capabilities, such as dynamic capabilities (Eisenhardt and Martin 2000; Sirmon et al. 2007) that help them overcome the disruptions and costs of structural change to enjoy the benefits? Are these capabilities present due to the maturity or life cycle of the firm?

Next, we turn to the studies on information processing and executive decision making to shed more light on this question of if/when firms may or may not reconfigure. Recall that literature on threat-rigidity finds that executives often revert back to path-dependent behavior in times of instability. Further reinforcing this behavior may be the centralization of decision making, narrowing of information gathered, faulty interpretations based on missing data, and a lack of perception that there is any threat that warrants structural change. Research on aspiration levels informs us that even if interpretations and perceptions are accurate, without enough discrepancy as measured against executives' aspirations, there may be no motivation or incentive to pursue change. These theories inform our current conflict. First, the case-study firms were highly decentralized, thus we infer that there was broad information gathered and more executives interpreting the data, increasing the likelihood that environmental threats would be perceived. Second, if the actively-changing firms did foster an environment of experimentation then executives may have had exceedingly high expectations (creating a significant difference between aspirations and performance) that further fostered structural change.

There are further implications from past literature that may shed light on our understanding of the different findings. If path-dependence is prevalent, is it that the case-study firms reconfigured in the past and this led to their reconfiguring again? Stated differently, did they have a proclivity toward structural change? Along with negative performance, studies have advocated that positive performance and growth may foster reconfiguration. Were these firms doing well? Is it important for us to compare their performance with rivals? After all, if they did reconfigure and their rivals did not, this may be a relevant distinction. There is much literature on the effects of the environment (i.e., markets) on structure. Are the industries within which the case-based firms are operating somehow different from those studied in the empirical papers? Do the industries where reconfiguration is observed have different levels of concentration, is competition more fast paced, are they in different stages of industry life cycle?

Finally, we question what is not observed by scholars. Though empirical work did not find the expected changes from environmental turbulence, we wonder whether there are changes inside firms that researchers did not observe; for example, were there changes in processes, routines, or rules? Further, if we observe a lack of reconfiguration in firms, is this indicative of a lack of ability or choice? If it is the latter, meaning that firms that correctly perceive threats and have the ability to make structural changes and yet choose not to, then, it raises an interesting question – when is the cumulative threat enough to initiate reconfiguration? These are but a few of the questions that warrant future study and that may inform theories of organization design.

5.4.2 *Informing Future Firms and the Practice of Organization Design*

There are many issues around ability, choice, and methods of structural change that will be particularly relevant in the future. Recall that Huber (2004) describes a future where firms are faced with increasing environmental dynamism (comprised of velocity, turbulence, and instability) and the need to make decisions sooner and more often.

Before firms can make the decision of whether to change their structures, they must first process information from both within and external to the firm to judge if they have the capability to initiate successful change. Thus, the design of structure to optimally gather information will be critical for firms in the future. Huber stresses the importance of sensing and interpreting the environment and notes that firms will be aided by advancements in information technology and better organization to process information. To achieve this, he proposes that future firms will more frequently use networks of teams and modular architecture. One can view reconfiguration as a means of accomplishing these new designs; by reorganizing business units under one umbrella there may be lower boundaries between units that foster interactions between employees, or efficiencies may be achieved from autonomy and specialization by modularizing units.

Another critical component that will affect decision making is the intent behind reconfiguration. For organizations to make the choice to disrupt the status quo and initiate structural change, they must first have a sense of what they hope to accomplish. One goal, as mentioned above, may be to separate units and further modularize them. However, a different goal may be to bring knowledge and resources together in new combinations by recombining units (Karim 2006). Huber mentions the importance of this as he predicts that future firms will need to innovate to survive and that this will require them to better manage (and potentially recombine) knowledge.

An important observation that Huber makes is that future firms will be more likely to buy knowledge than to internally develop it and will more frequently use structural arrangements to acquire and integrate this knowledge from other firms. We believe that the concept of reconfiguration addresses these potential structural arrangements as firms decide how to organize targets that were recently acquired – do they leave them autonomous or integrate them with other business units? These are several of the means by which structural reconfiguration addresses the organization design of firms in the future.

Acknowledgments The authors wish to thank Anne Bøllingtoft, George Huber, and participants in the 2008 Aarhus Conference on Organization Design for their helpful comments.

References

Abernathy W, Utterback J (1978) Patterns of Industrial Innovation. Technology Review 80: 40–47.

Amburgey TL, Kelly D, Barnett WP (1993) Resetting the Clock: The Dynamics of Organizational Failure. Administrative Science Quarterly 38(1): 51–73.

Baldwin CY, Clark KB (2000) Design Rules: The Power of Modularity. Boston: MIT Press.

Bergh DD (1998) Product-Market Uncertainty, Portfolio Restructuring, and Performance: An Information-Processing and Resource-Based View. Journal of Management 24 (2): 135–155.

Blau PM (1970) A Formal Theory of Differentiation in Organizations. American Sociological Review 35(2): 201–218.

Bowman EH, Singh H (1993) Corporate Restructuring: Reconfiguring the Firm. Strategic Management Journal 14 (Special Issue: Corporate Restructuring): 5–14.

Burns T, Stalker GM (1961) The Management of Innovation. London: Tavistock.

Burton RM, Lauridsen J, Obel B (2002) Return on Assets Loss from Situational and Contingency Misfits. Management Science 48 (11): 1461–1485.

Burton RM, Obel B (2004) Strategic Organization Diagnosis and Design: The Dynamics of Fit, 3rd edn. Dordrecht: Kluwer Academic Publishers.

Cafferata GL (1982) The Building of Democratic Organizations: An Embryonic Metaphor. Administrative Science Quarterly 27: 280–303.

Chandler A (1962) Strategy and Structure: Chapters in the History of the American Industrial Enterprise. Cambridge, MA: MIT Press.

Chattopadhyay P, Glick WH, Huber GP (2001) Organizational Actions in Response to Threats and Opportunities. Academy of Management Journal, 44 (5): 937–955.

Chesbrough H, Teece D (1996) When is Virtual Virtuous: Organizing for Innovation. Harvard Business Review, 74(1): 65–74.

Child J (1972) Organizational Structure, Environment and Performance: The Role of Strategic Choice. Sociology 6 (1): 1–22.

Child J, Kieser A (1981) Development of Organizations over Time. In: Starbuck W, Nystrom P (eds), Handbook of Organization Design. New York: Oxford, pp 28–64.

Ciborra CU (1996) The Platform Organization: Recombining Strategies, Structures, and Surprises. Organization Science 7 (2): 103–118.

Crozier M (1964) The Bureaucratic Phenomenon. Chicago: University of Chicago Press.

Cyert RM, March JG (1963) A Behavioral Theory of the Firm. Cambridge: Blackwell Publishing.

Dalton DR, Daily CM, Ellstrand AE, Johnson JL (1998) Meta-Analytic Reviews of Board Composition, Leadership Structure, and Financial Performance. Strategic Management Journal 19: 269–290.

Donaldson L (1985) Organization Design and the Life Cycles of Products. Journal of Management Studies 22 (1): 25–37.

Donaldson L (1987) Strategy and Structural Adjustment to Regain Fit and Performance: In Defense of Contingency Theory. Journal of Management Studies, 24 (1): 1–24.

Donaldson L (1990) The Normal Science of Structural Contingency Theory. In: Clegg H, Nord W(eds), The Handbook of Organization Studies. London: Sage.

Donaldson L (1995) American Anti-Management Theories of Organization: A Critique of Paradigm Proliferation. New York: Cambridge University Press.

Donaldson L (2001) The Contingency Theory of Organizations. Thousand Oaks, CA: Sage.

Eisenhardt KM, Brown SL (1999) Patching: Re-Stitching Business Portfolios in Dynamic Markets. Harvard Business Review 77(3): 71–82.

Eisenhardt KM, Martin JA (2000) Dynamic Capabilities: What are They? Strategic Management Journal, Special Issue 21(10–11): 1105–1121.

Feldman MS, Pentland BT (2003) Organizational Routines as a Source of Flexibility and Change. Administrative Science Quarterly 48(1): 94–118.

Galunic DC, Eisenhardt KM (1996) The Evolution of Intracorporate Domains: Divisional Charter Losses in High-Technology, Multidivisional Corporations. Organization Science 7(3): 255–282.
Galunic DC, Eisenhardt KM (2001) Architectural Innovation and Modular Corporate Forms. Academy of Management Journal 44 (6): 1229–1249.
Garud R, Kumaraswamy A (1995) Technological and Organizational Designs for Realizing Economies of Substitution. Strategic Management Journal 16 (Special Issue: Technological Transformation and the New Competitive Landscape): 93–109.
Greiner L (1972) Evolution and Revolution as Organizations Grow. Harvard Business Review 50(4): 37–46.
Gresov C (1989) Exploring Fit and Misfit with Multiple Contingencies. Administrative Science Quarterly 34(3): 431–454.
Greve HR (2003) Organizational Learning from Performance Feedback: A Behavioral Perspective on Innovation and Change. Cambridge: Cambridge University Press.
Hage J, Aiken M (1969) Routine Technology, Social Structure and Organizational Goals. Administrative Science Quarterly 14(3): 366–376.
Hambrick DC, Mason PA (1984) Upper Echelons: The Organization as a Reflection of its Top Managers. Academy of Management Review 9: 193–206.
Hannan MT, Freeman J (1984) Structural Inertia and Organizational Change. American Sociological Review 49: 149–164.
Haveman H (1992) Between a Rock and a Hard Place: Organizational Change and Performance under Conditions of Fundamental Environmental Transformation. Administrative Science Quarterly 37: 48–75.
Helfat CE, Eisenhardt KM (2004) Inter-Temporal Economies of Scope, Organizational Modularity, and the Dynamics of Diversification. Strategic Management Journal 25(13): 1217–1232.
Henderson RM, Clark KB (1990) Architectural Innovation: The Reconfiguration of Existing Product Technologies and the Failure of Established Firms. Administrative Science Quarterly 35(1): 9–30.
Hill CWL, Hitt MA, Hoskisson RE (1992) Cooperative versus Competitive Structures in Related and Unrelated Diversified Firms. Organization Science 3(4): 501–521.
Hirsch PM, De Soucey M (2006) Organizational Restructuring and its Consequences: Rhetorical and Structural. Annual Review of Sociology 32: 171–89.
Hoskisson RE, Johnson RA (1992) Corporate Restructuring and Strategic Change: The Effect on Diversification Strategy and R&D Intensity. Strategic Management Journal 13 (8): 625–634.
Hoskisson RE, Johnson RA, Moesel DD (1994) Corrporate Divestiture Intensity in Restructuring Firms: Effects of Governance, Strategy, and Performance. Academy of Management Journal 37(5): 1207–1251.
Hoskisson RE, Turk TA (1990) Corporate Restructuring: Governance and Control Limits of the Internal Capital Market. Academy of Management Review 15 (3): 459–477.
Huber GP (2004) The Necessary Nature of Future Firms: Attributes of Survivors in a Changing World. Boston, MA: Sage.
Jensen MC (1986) Agency Costs of Free Cash Flow. American Economic Review 76: 323–329
Jensen MC (1988) Takeovers: Their Causes and Consequences. Journal of Economic Perspectives 2 (1): 21–48.
Johnson R (1996) Antecedents and Outcomes of Corporate Refocusing. Journal of Management 22 (3): 439–483.
Karim S (2006) Modularity in Organizational Structure: The Reconfiguration of Internally Developed and Acquired Business Units. Strategic Management Journal 27: 799–823.
Karim S, Mitchell W (2004) Innovating Through Acquisition and Internal Development. Long Range Planning 37: 525–547.
Keats BW, Hitt MA (1988) A Causal Model of Linkages Among Environmental Dimensions, Macro Organizational Characteristics, and Performance. Academy of Management Journal 31 (3): 570–598.

Khandwalla PN (1977) Organizational Design. New York: Harcourt, Brace, Jovanovich.
Kimberley J, Miles RH (1980) The Organizational Life Cycle. San Francisco: Jossey Bass.
Lant TK, Hurley AE (1999) A Contingency Model of Response to Performance Feedback: Escalation of Commitment and Incremental Adaptation in Resource Investment Decisions. Group and Organization Management 24 (4): 421–437.
Lawrence PR, Lorsch JW (1967) Organization and Environment: Managing Differentiation and Integration. Boston: Harvard Business School.
Lei D, Hitt MA, Goldhar JD (1996) Advanced Manufacturing Technology: Organizational Design and Strategic Flexibility. Organization Studies 17 (3): 501–524.
Liebeskind J, Wiersema M, Hansen G (1992) LBOs, Corporate Restructuring, and the Incentive-Intensity Hypothesis. Financial Management 21(1): 73–88.
Lopes LL (1987) Between Hope and Fear: The Psychology of Risk. Advances in Experimental Social Psychology 20: 255–295.
March JG (1981) Footnotes to Organizational Change. Administrative Science Quarterly 26: 563–577.
March JG, Shapira Z (1992) Variable Risk Preferences and the Focus of Attention. Psychological Review 99 (1): 172–183.
Merton RK (1957) Social Theory and Social Structure, Rev. ed. Glencoe, IL: Free Press.
Miller D (1988) Relating Porter's Business Strategies to Environment and Structure: Analysis and Performance Implications. Academy of Management Journal 31 (2): 280–308.
Mintzberg H (1980) Structure in 5s: A Synthesis of the Research on Organization Design. Management Science 26 (3): 322–341.
Nadler D, Tushman ML (1984) A Congruence Model for Diagnosing Organizational Behavior. In: Kolb DA, Rubin JM, McIntyre JM (eds), Organizational Psychology: Readings on Human Behavior in Organizations. Englewood Cliffs, NJ: Prentice-Hall, pp 587–603.
Nadler DA, Tushman M (1997) Competing by Design: The Power of Organizational Architecture. New York: Oxford University Press.
Naman JL, Slevin DP (1993) Entrepreneurship and the Concept of Fit: A Model and Empirical Tests. Strategic Management Journal 14 (2): 137–153.
Ocasio WC (1995) The Enactment of Economic Adversity: A Reconciliation of Theories of Failure-Induced Change and Threat-Rigidity. In: Staw BM, Cummings LL (eds), Research in Organization Behavior Greenwich, CT: JAI Press, 17: 287–331.
Penrose ET (1959) The Theory of the Growth of the Firm. New York: Wiley.
Porter ME (1987) From Competitive Advantage to Corporate Strategy. Harvard Business Review 65 (3): 43–59.
Phan PH, Hill CWL (1995) Organizational Restructuring and Economic Performance in Leveraged Buyouts: An Ex Post Study. Academy of Management Journal 38 (3): 704–739.
Quinn RE, Cameron K (1983) Organization Life Cycles and Shifting Criteria of Effectiveness: Some Preliminary Evidence. Management Science 29 (1): 33–51.
Ravenscraft DJ, Scherer FM (1987) Mergers, Sell-offs, and Economic Efficiency. Washington, DC: Brookings Institution Press.
Romanelli E, Tushman ML (1994) Organization Transformation as Punctuated Equilibrium: An Empirical Test. Academy of Management Journal 34: 141–1166.
Rumelt RP (1974) Strategy, Structure and Economic Performance. Boston, MA: Harvard University Press, Cambridge.
Sanchez R, Mahoney JT (1996) Modularity, Flexibility, and Knowledge Management in Product and Organization Design. Strategic Management Journal 17: 63–76.
Schilling MA (2000) Toward a General Modular Systems Theory and Its Application to Interfirm Product Modularity. Academy of Management Review 25 (2): 312–334.
Schilling MA, Steensma HK (2001) The Use of Modular Organizational Forms: An Industry-Level Analysis. Academy of Management Journal 44 (6): 1149–1168.
Seth A, Easterwood J (1993) Strategic Redirection in Large Management Buyouts: The Evidence from Post-Buyout Restructuring. Strategic Management Journal 14 (4): 251–273.

Sirmon DG, Hitt MA, Ireland RD (2007) Managing Firm Resources in Dynamic Environments to Create Value: Looking Inside the Black Box. Academy of Management Review 32 (1): 273–292.

Starbuck W, Milliken F (1988) Executives' Perceptual Filters: What They Notice and How They Make Sense. In: Hambrick D (eds), The Executive Effect: Concepts and Methods for Studying Top Executives. Greenwich, CT: JAI Press, pp 35–65.

Staw BM, Ross J (1987) Understanding Escalation Situations: Antecedents, Prototypes, and Solutions. Research in Organizational Behavior 9: 39–78.

Staw BM, Sandelands LE, Dutton JE (1981) Threat-Rigidity Effects in Organizational Behavior: A Multilevel Analysis. Administrative Science Quarterly 26: 501–524.

Teece DJ, Pisano G, Shuen A (1997) Dynamic Capabilities and Strategic Management. Strategic Management Journal 18 (7): 509–533.

Thomas J, Clark S, Gioia D (1993) Strategic Sensemaking and Organizational Performance: Linkages among Scanning, Interpretation, Action, and Outcomes. Academy of Management Journal 36: 239–270.

Tushman ML, Murmann JP (1998) Dominant Designs, Technology Cycles, and Organizational Outcomes. In: Staw B, Sutton R (eds), Research in Organizational Behavior. Greenwich, CT: JAI Press, Vol. 20.

Utterback JM (1994) Mastering the Dynamics of Innovation. Boston, MA: Harvard Business School Press.

Utterback JM, Suárez F (1993) Innovation, Competition, and Industry Structure. Research Policy 22: 1–21.

Van de Ven AH, Poole MS (1995) Explaining Development and Change in Organizations. Academy of Management Review 20 (3): 510–540.

Walsh JP (1988) Top Management Turnover Following Mergers and Acquisitions. Strategic Management Journal 9: 173–183.

Walsh JP (1989) Doing a Deal: Merger and Acquisition Negotiations and Their Impact upon Target Company Top Management Turnover. Strategic Management Journal 10: 307–322.

Walsh JP, Ellwood JW (1991) Mergers, Acquisitions, and the Pruning of Managerial Deadwood. Strategic Management Journal 12 (3): 201–217.

Zenger TR, Hesterly WS (1997) The Disaggregation of Corporations: Selective Intervention, High-Powered Incentives, and Molecular Units. Organization Science 8 (3): 209–222.

Chapter 6
Embedding Virtuality into Organization Design Theory: Virtuality and Its Information Processing Consequences

Kent Wickstrøm Jensen, Dorthe Døjbak Håkonsson, Richard M. Burton, and Børge Obel

Abstract What is virtuality in organizational design? In this chapter we argue for the importance of understanding the nature and effect of the characteristics of virtual organizations, rather than simply focusing on how these characteristics are different from co-located organizations. Through a review of literature relating to virtual organizations we identify two different dimensions: *locational* and *relational* differentiation, which capture the nature of virtual organizations well. We anchor theoretically these dimensions to organization design and information processing theory. This enables us to identify their effects and consequences for coordination in information processing terms. We thereby not only integrate theory of virtual organization into extant theory of organization design but, more importantly, also demonstrate how increasing virtuality essentially imposes an information processing dilemma for organizations: Locational differentiation reduces the information processing capacity, while relational differentiation increases the information processing requirements. We discuss the managerial as well as the theoretical implications of these findings.

Keywords Virtual organizations · Virtuality · Coordination · Information processing theory

6.1 Introduction

The attractiveness of virtual organizational designs is widely recognized in the organization literature. In particular, they are known for their capacity to exploit diverse sets of geographically and intra-organizationally dispersed capabilities to provide for flexible and fast responses to changing market conditions (Mowshowitz 1994; Fulk and DeSanctis 1995; Ahuja and Carley 1999). For those reasons, they seem

K. Wickstrøm Jensen (✉)
Department of Entrepreneurship and Relationship Management, University of Southern Denmark, Engstien 1, 6000, Kolding, Denmark
e-mail: kwj@sam.sdu.dk

A. Bøllingtoft et al. (eds.), *New Approaches to Organization Design*, Information and Organization Design Series 8, DOI 10.1007/978-1-4419-0627-4_6,

to contain promising design elements for organizations of the future. Similarly, and contrary to these promises, it is just as widely recognized that virtual organization design imposes particular challenges in terms of coordinating across time, space, and organizational boundaries (Lipnack and Stamps 1997).

Abundant reviews have served to describe what virtual organizations or virtuality is, identifying the characteristics of virtuality that at once provide for flexibility and speed but also impose significant coordination challenges for the organization (Martins et al. 2004; Kirkman and Mathieu 2005; Chudoba et al. 2005). Much less literature has been concerned with describing what the implications of virtuality are – particularly in terms of the coordination requirements it poses on organizations.

Early writings on virtuality focused on describing their specific characteristics, as distinct from those of traditional type organizations (e.g., Guzzo and Dickson 1996). Thus, virtual organizations, as opposed to co-located or traditional organizations, rely on technology-mediated communication. Further, they are often described to cross boundaries of time, geography, and organizations (Lipnack and Stamps 1997).

In contrast, recent contributions have imposed a more continuous perspective on virtual organizing. For instance, Chudoba et al. (2005), Griffith et al. (2003) and Griffith and Neale (2001) have argued that the distinction between co-located and virtual organizations is artificial. Instead, they argue that we should focus on describing organizations by their degree of virtuality. As an example of this perspective, Kirkman and Mathieu (2005: 702) define team virtuality as "(a) the extent to which team members use virtual tools to coordinate and execute team processes, (b) the amount of informational value provided by such tools, and (c) the synchronicity of team member virtual interaction." In much the same way, Chudoba et al. (2005) put up a set of measures of virtual characteristics that would lead to a decrease in cohesion. In this way, many of the later contributions relate to how individual characteristics impact coordination difficulties in virtual organizations in various ways. Yet, they focus less on integrating these different views and perspectives into a common framework to enhance our overall understanding of their impact.

We agree with recent contributions that it makes sense to see virtuality as degree rather than as a distinct form. Indeed, virtual characteristics can be identified with a varying degree in more or less all organizations and at various levels within organizations. As such, while we in this chapter focus primarily on virtuality at the organizational level, it applies equally well to teams and other organizations.

Yet, even more important, the focus should be shifted toward understanding the nature and effects of virtual characteristics, thereby moving toward a better understanding of the functioning of highly virtual organizations, ultimately enabling us to support their coordination needs more effectively through design.

Through a review of literature we illuminate the differential nature and relationships between the characteristics used to define organizations as virtual. We identify two dimensions which seem to capture their diversity well. To advance the understanding of the effect on coordination of these dimensions, we anchor the dimensions to information processing and organizational design theory. This enables us to identify their effects and consequences for information processing in an organization.

Conceptualizing virtuality in terms of its effect on information processing requirements and capacity respectively, not only advances our understanding of virtual design by itself, but also provides a more theoretically grounded avenue for understanding how virtuality relates to other organization design strategies (Galbraith 1973) and other organization design parameters such as centralization and formalization (Burton and Obel 2004).

6.2 Theoretical Background

Following Oxford English Dictionary, virtual is defined as something which is opposed to the actual, that is, "A virtual (as opposed to an actual) thing, capacity, etc.; a potentiality" or alternatively as "Essential nature or being, apart from external form or embodiment."

Corresponding to this dichotomous distinction between virtual and actual, the first developments of virtuality in the organization literature proposed a view of the virtual organization as a distinct archetype to be identified on a set of structural characteristics that in some way depart from traditional organizational structures. Davidow and Malone (1992) were among the first to discuss the concept of virtuality in relation to organizations. Thus, inspired by the potential of new information technology to delimit previous design types, they defined the virtual organization as: "a group of independent firms which link together and form a single temporary company[1]." Since this early proposition, much has been done to identify the characteristics of virtual organizations that depart from the "actual."

The majority of this research converges toward definitions of virtual organizations as spatially separated organizational units relying on information technology to mediate the communication and coordination processes. In this chapter, we concur with DeSanctis and Monge's definition: "a collection of geographically distributed, functionally and/or culturally diverse entities that are linked by electronic forms of communication and rely on lateral, dynamic relationships for coordination. Despite its diffuse nature, a common identity holds the organization together in the minds of members, customers, or other constituents" (1999: 693). DeSanctis and Monge's definition is encompassing in the sense that it incorporates the traditional dimensions of virtual organizations, while at the same time is not restricted to focusing only on issues of virtual vs. co-located notions of virtuality.

Alternatively, Martins et al. define virtual teams as (2004: 808) "teams whose members use technology to varying degrees in working across locational, temporal, and relational boundaries to accomplish an interdependent task." Locational boundary refers to "any physical dispersion of team members, such as different

[1] While the literature discusses both virtual teams and virtual organizations, in this chapter we refer to both as virtual organizations. Even though teams are likely to be more transient than organizations, this is not a distinction that matters for the purposes of this chapter, since we are focusing on conceptualizing the information processing characteristics of virtuality, per se.

geographic locations or different workplaces at the same geographic location." The temporal boundary (2004: 808) "encompasses lifecycle and synchronicity." The relational boundary refers to (2004: 808): "the differences in relational networks of VT members, that is, their affiliation with other teams, departments, organizations, and cultural sub-groups." As highlighted by Martins et al. (2004) this type of definition moves away from describing virtual organizing solely in terms of their "virtual-ness," but also focuses on their degree of "team-ness." By adding a notion of "team-ness," we believe they also point toward the importance of communication and coordination across virtual organizations. As argued by these authors their addition encourages "a focus on understanding the functioning of VTs rather than on simply comparing them to face-to-face teams." As argued, we wish to focus on the coordination issues of virtual organizations to further an understanding of this particular aspect of their functioning.

That coordination is essential is also indicated in DeSanctis and Monge's definition, since virtual organizations face an extraordinary challenge in achieving effective coordination among their geographically, functionally, and/or culturally separated, and highly interdependent entities.

We believe it is worthwhile to push an encompassing understanding of the aspect of virtual functioning that relates to coordination. Many studies have examined aspects of how various characteristics of virtual organization impact coordination, thereby contributing toward a better understanding of the functioning of virtual organizations. In the following review, we will argue that these characteristics and their associated impact on coordination can be captured by two overall dimensions. While this way building on work that is both related to older and newer ways of defining virtual organizing, our focus is on a better understanding of the functioning of these organizations, particularly with regards to their coordination requirements.

6.2.1 Dimensions of Virtual Organizations and Their Consequences for Coordination

Probably one of the most highlighted characteristics of virtual organizations is their spatial differentiation. As early as 1977, Allen studied how distances between people negatively affected their interactions and showed that the frequency of work-related technical communication between co-workers drops rapidly as distance between them increases, and further that communication frequency reaches an asymptote as the distance approaches 30 m.

More recently, Kraut et al. (2002) studied the larger consequences of spatial separation, and argued that spatial separation has detrimental effects on the likelihood as well as the quality of communication, mainly because it disables face-to-face communication. They provide a review of how current computer-mediated communication technologies' visibility, co-presence, mobility, co-temporarity and other affordances affect the important collaborative tasks of initiating conversation, establishing common ground, and maintaining awareness of potentially relevant changes in the collaborative environment. Based on that review, they argue that computer-mediated communication techniques do not provide the interpersonal interactions and awareness that physical proximity does. Thus, even though proximity by itself is not a guarantee for interactions, and that organizational members, even if they are proximate, may never meet, the physical dispersion of virtual teams has been argued to decrease the likely effectiveness of communication, mainly because it disables face-to-face communication. Similar arguments were made by O'Leary and Cummings (2007), who argued that spatial dispersion decreased the likelihood of face-to-face interaction.

Closely related to spatial separation is the temporal separation. Often, geographical distances will be associated with time differences among organizational units, making work hours less likely to overlap. Even though this dimension is related to spatial separation, the relationship, of course, is not linear. Spatial dispersion between east-to-west will likely entail fewer work hour overlaps, whereas organizations dispersed on a north-to-south axis will have more work hour overlaps. In terms of coordination difficulties, temporal separation makes synchronous communication less likely, that is, it reduces real-time problem solving (Grinter et al. 1999; Herbsleb et al. 2000; Malone and Crowston 1994). Even if computer-mediated technologies such as e-mail are relied on, temporal separation may still create serious delays in communications as members in one geographical location may have to await decisions from members in another physical location (Kumar 2006; O'Leary and Cummings 2007).

While spatial and temporal separation are not the same and also may not have a linear relationship, the coordination consequences of the two characteristics are still similar in the sense that research reveals how their main effect relates to that of disabling face-to-face communication, and the potentially detrimental consequences of this. By this, we believe they both relate to what Martins et al. (2004) referred to as a "locational boundary." As mentioned, Martin et al. defined locational boundaries as a boundary that refers to "any physical dispersion of team members, such as different geographic locations or different workplaces at the same geographic location" (2004: 808). Martins et al. did not, however, elaborate on the further implications of these dimensions, but used them only for proposing an integrative definition of virtual organizations. Because both spatial and temporal separation relate to the disability to use certain communication media among organizational units, as distinct from Martins et al., we will relate to them in the following as locational differentiation.

Normally, when speaking of differentiation in an organizational design context, the concern is about horizontal and vertical differentiation respectively; horizontal differentiation referring to specialization by education, experience, training, and tasks; and vertical differentiation referring to the number of levels in the hierarchy (e.g., Burton et al. 2006).

As mentioned above, one of the motivations behind virtual organizing is to capitalize on experts displaced across wide geographical areas and across organizational boundaries. Therefore, virtual organizing can by purpose be expected to exhibit a high horizontal differentiation.

As noted by Lawrence and Lorsch (1967), however, differentiation is not restricted simply to the division of work, but may encompass any element that differentiates between behavioral attributes of organizational members. We believe it is reasonable to talk about both temporal and spatial separation as having the potential to differentiate between behavioral attributes of organizational members. In the following, we will therefore relate to the two as locational differentiation, to emphasize how their joint effect serves to differentiate between members in an organization by disabling the use of certain communication media.

Further, a range of other characteristics have been proposed, which seem to relate to a different aspect of coordination in virtual organizations. We will refer to these as cultural, temporal, functional, and inter-organizational differentiation, respectively.

Relating to cultural differentiation, many studies have examined how national as well as organizational cultural differences may hinder building shared organizational cultures (Horii et al. 2005; Chesbrough and Teece 1996). While these differences are also likely to be at least non-linearly correlated to spatial differentiation, their impact on coordination is different in the sense that their impact relates more to the ambiguities they create. Without shared cultures, inefficiencies in communication are likely to arise due to lack of common values, assumptions, or perceptions. For instance, studies have shown how different cultures may hinder the development of trust in virtual organizations (Frey and Schlosser 1993; Handy 1995). In that sense, cultural differentiation refers to differences in values, attitudes, customs, etc., arising as consequences of dispersed organizational units. These differences may, or may not, derive from differences in national culture. What is further important to notice here is that there is no direct relationship between spatial and cultural differentiation although these are positively correlated. As an example, a distance of 2000 km within the United States likely involves considerable less cultural differences than the same distance would entail when spanning countries like Germany, Croatia, Ukraine, and Turkey.

Relating to temporary differentiation, virtual organizations often allow temporary teams to form in order to solve and exploit emerging opportunities. When the temporary differentiation is high, members have only little or no previous experience of working together. In such situations, transactive memory is low, trust may be low, roles are undefined or under negotiation, and routines for decision-making and coordination are fairly under developed. This may well increase the difficulty of sharing knowledge and information (Reagans and McEvily 2003). While both of the latter effects are also known from co-located organizations, the effect is different in virtual organizations. In co-located organizations, employees have the opportunity to meet face-to-face, which has been demonstrated to facilitate the creation of trust. Hence, building trust is difficult when employees are not co-located, when media are impersonal, and when working relationships are temporary and of short length duration.

Concerning functional differentiation, virtual organizations are often made up of experts located in diverse places (Alavi 1993; Finholt et al. 1990), and virtual organizations therefore tend to be more specialized than co-located organizations. Deep expertise (Simon 1991; Mowshowitz 1997) can reduce the uncertainty and equivocality in localized information that otherwise would perpetuate through the organization. This naturally implies that coordination and their contribution to performance becomes essential for these organizations' survival. Coordination of highly specialized experts in different domains, however, is not a trivial matter. Reaching agreement among experts may be troublesome due to their different cognitive orientations and lack of mutual knowledge (Rogers and Bhowmik 1970; Cohen and Levinthal 1990; Cramton 2001), distrust due to social categorization (Brewer 1996; Kramer 1999), and differences in cultural norms, values, and attitudes (Armstrong and Cole 2002). Thus, increased functional specialization may similarly lead to an increased coordination load.

Much work on virtual organization describes boundaries of virtual organizations as edgeless and permeable (Mowshowitz 1994; Nohria and Berkley 1994). Inter-organizational relationships are continually redefined relative to changes in the component structure of products and technologies and relative to changes in the customer base (Monge and DeSanctis 1999). Inter-organizational arrangements share a high potential for various kinds of differentiation including cultural and functional differentiation. Yet, among the particular kinds of characteristics for coordination across organizational boundaries are differences in management styles, reporting structures, conflict resolution methods, and other coordination and decision-making routines. We will refer to these differences in terms of inter-organizational differentiation.

Overall, these latter four characteristics, while they may be related to spatial separation, clearly have a different effect on coordination in the sense that all studies highlight how ambiguities occur as differentiation increases on these characteristics. In that sense, they seem to relate to the characteristic "relational boundary" which Martins et al. (2004: 808) used in their definition of virtual teams, that is, the "differences in relational networks of VT members, that is, their affiliations with other teams, departments, organizations, and cultural sub-groups." Further, they also seem to capture what Martins et al. (2004) referred to as lack of "team-ness" in virtual organizations. As with locational differentiation, we refer to these in the following as relational differentiation because they also encompass elements that differentiate between behavioral attributes of organizational members (Lawrence and Lorsch 1967). Yet, as distinct from locational differentiation, relational differentiation and its four underlying dimensions all describe different potentials for differences in cognitive orientations among members of organizational units which will likely influence coordination.

Focusing on the coordination requirements imposed by previously well-known characteristics of virtual organizations, we have proposed that these seem to be captured well by two overall dimensions. We refer to these using the notions of Martins et al. (2004): locational and relational differentiation. These two dimensions, along with their underlying characteristics are illustrated in Fig. 6.1. As also appears from Fig. 6.1, we argue that locational differentiation impacts on coordination mainly in terms of restricting the use of certain communication media. Relational differentiation in contrast, seems to impact coordination by increasing ambiguities in communication. Finally, as illustrated by the dotted arrow in Fig. 6.1., we recognize

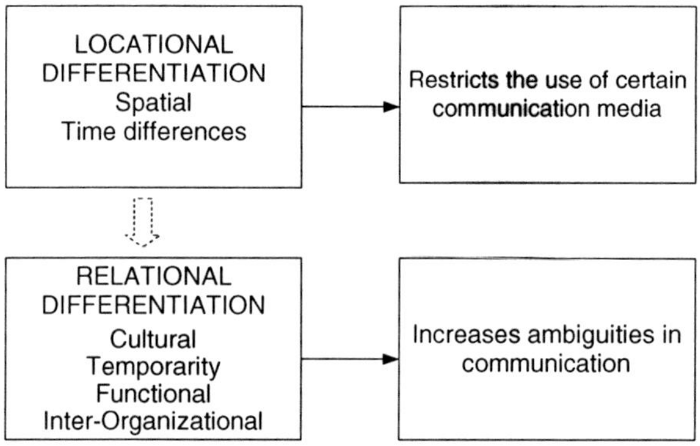

Fig. 6.1 Locational and relational differentiation and their consequences for coordination

the potential effect of locational differentiation on relational differentiation, yet maintain our focus on the separate impacts of each of these two dimensions on coordination needs.

Overall, we believe this way of representing previous literature is useful in itself in that it points toward the differential coordination needs underlying virtual organizations. Still, we believe we also need to move toward a better understanding of the impact of these differential coordination needs.

6.2.2 Toward a Deeper Understanding of the Coordination Needs of Virtual Organizations

Recent writings have proposed similar ideas as those presented in Fig. 6.1. For instance, O'Leary and Cummings (2007) emphasized the importance of focusing on the differential outcomes that characteristics of virtual organizations may have. They characterized the differential outcomes that spatial, temporal, and configurational dispersion would have on the likelihood of face-to-face communication, on synchronous interaction, and on dependencies, remoteness and negative subgroups dynamics, respectively. As appears, their description of discontinuities is similar to our notion of relational differentiation, except that we argue for a different effect of spatial differentiation, which is therefore not included in our notion of relational differentiation.

Further, even though each of these outcomes are more or less related to communication difficulties just as they seem related to our above arguments, they do not, however, provide an understanding of the implications of these effects.

Chudoba et al. (2005) in some way also relate to the differential outcomes of various dimensions of virtuality. More precisely, Chudoba et al. (2005) propose to compile previous conceptualizations of virtuality into a unified construct which Watson-Manheim et al. (2002) introduced as "discontinuities" defined as "factors contributing to a decrease in cohesion" (2005: 280). Following these authors, discontinuities could be physical location, time zone differences, and cultural differences. While these are separable, they also often come in bundles (e.g., time zone and physical location differences). This seems a fruitful way forward in terms of focusing on how dimensions, either by themselves or together, group into categories that can be described by their impact on particular outcomes. Even though their outcome measure is cohesion, it is not difficult to imagine that coordination costs would increase along with a decrease in cohesion. Yet, similar to the O'Leary and Cummings findings, this understanding nevertheless does not bring us toward an understanding of the overall implications, or underlying mechanisms of these coordination needs.

One paper which seems to initiate such an understanding is Kirkman and Mathieu (2005). They focus on IT as a defining characteristic of virtual organizations and propose a multidimensional view on virtuality. As also mentioned, they define virtuality by the three dimensions of: (a) the extent to which team members use virtual tools to coordinate and execute team processes, (b) the amount of informational value provided by such tools, and (c) the synchronicity of team member virtual interaction. To our purposes, we find the notion of "informational value" fruitful. Information value is derived from the concept of media richness (Daft and Lengel 1986; Venkatesh and Johnson 2002) and relates to "the extent to which virtual tools send or receive communication or data that is valuable for team effectiveness" (2005: 703). What is interesting about this notion is how the introduction of the concept of information value introduces a possible measure for not only capturing the essence, but also moving toward capturing the implications of the various measures. Kirkman and Mathieu, however, only use information processing theory to define their informational value dimension. Furthermore, they focus on IT as a defining characteristic, thereby adopting a more restricted and to some extent more controversial view on virtual organizations than many previous authors.

Nevertheless, in terms of moving toward an encompassing understanding of the differential coordination requirements inherent in virtual organizations, we believe it makes sense to anchor the various dimensions of virtuality into a consistent theoretical framework. This would not only allow a consistency in our description of the various dimensions, but more importantly also allow a comparison of their differential requirements as well as their eventual interdependencies. Further, the information processing perspective seems relevant for our purpose, as it has as its main emphasis on coordination across organizations.

In the following section, we propose to build on information processing theory to enable a new conceptualization of virtuality which provides theoretical clarity in terms of describing the functioning of virtual organizations.

6.3 Virtualness in an Information Processing Perspective

When taking an information processing perspective on firms, an organization can be defined as an entity that uses information in order to coordinate and control its activities (Galbraith 1974; Arrow 1974). Building on information processing arguments, essentially an organization is well designed when its information processing capacity matches its information processing load (Galbraith 1974; Tushman and Nadler 1978). If an organization's information processing load is higher than the organizational information processing capacity, the organization will be ineffective and unable to reach organizational goals. On the other hand, if information processing capacity exceeds information processing load, the organization will consume superfluous resources and be inefficient.

According to Galbraith as well as Daft and Lengel (1986) the information processing load imposed on organizational coordination can be conceptualized in terms of uncertainty and equivocality.

Uncertainty relates to the absence of specific objective information to answer specific questions. Or, as defined by Galbraith (1973/1974), "the amount of uncertainty is a measure of the difference between the amount of information required to perform the task, and the amount of information already possessed." Equivocality, on the other hand, relates to the ambiguity and confusion arising around multiple and conflicting interpretations about an organization's situation. The higher uncertainty and/or equivocality an organization experiences, the higher its information processing requirements.

As argued by Galbraith (1974: 10) "the greater the uncertainty of the task, the greater the amount of information that has to be processed between decision makers." The prevalent view is that an organization's design will enable an organization to process the information required to reduce uncertainty. In order to retain this "fit" between information processing capacity and need, Galbraith (1974) proposes four generic information processing strategies: Organizations may lower the need for information processing through the creation of slack resources, or through the creation of self-contained tasks. Or, the organization can increase the information processing capacity through investments in vertical information systems, or by creating lateral relations.

Daft and Lengel (1986) further argue for a conceptual distinction between the structural mechanisms required to handle uncertain and equivocal information respectively. In situations of high equivocality, that is, "ambiguity, the existence of multiple and conflicting interpretations about an organizational situation" (1986: 555), structural mechanisms would have to enable clarification, and enactment more than simply provide large amounts of data. Hence, they argued that the key factor in equivocality reduction "is the extent to which structural mechanisms facilitate the processing of rich information" (1986: 559–560), where information richness is defined as the ability of information to change understanding within a time interval. In that sense, they argue: "information richness pertains to the learning capacity of a communication."

According to Daft and Lengel, the information processing load arises out of three sources: (1) Technology as defined along Perrow's two dimensions of task analyzability and task variety (Perrow 1967), (2) Interdepartmental relations as defined in terms of differentiation (Lawrence and Lorsch 1967) and task interdependence (Thompson 1967), and (3) Environmental forces in terms of equivocality, hostility and dynamics. Thus information processing need is not exclusively determined by

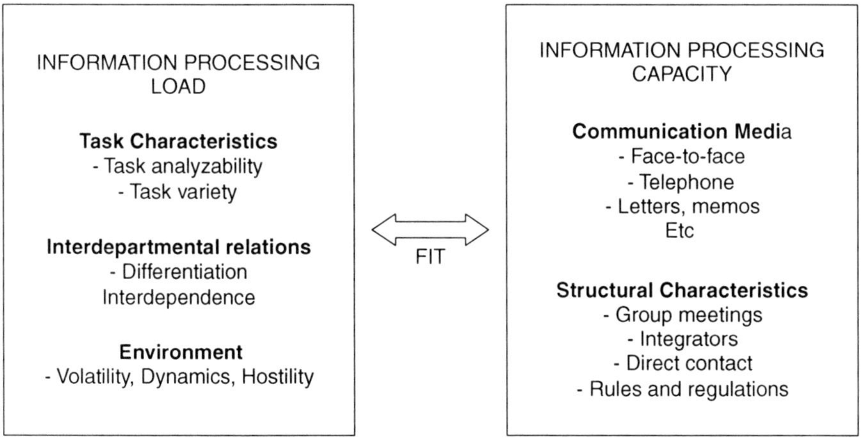

Fig. 6.2 Information processing load and capacity (Source: Daft and Lengel (1986))

environmental forces, but also by the way tasks are organized within the organization. This view is similar to Tushman and Nadler's (1978) discussion of external and internal sources of uncertainty.

As an extension to Galbraith's (1974) structural design variables, Daft and Lengel (1986) further emphasize the influence of communication technology on information processing capacity. Their essential idea is that there must be congruence between information richness and media richness. Information processing capacity is thus a function of the media chosen for the given communication, and particularly how well the media allows for immediate feedback, high number of cues and more channels utilized, personalization, and high language variety (Daft and Wiginton 1979).

Figure 6.2 summarizes the above discussions on information processing load and capacity by visualizing the need for a fit between the information processing capacity of an organization (i.e., its communication media and structural characteristics) and its organization's information processing load.

Relative to the objective of introducing the information processing view on virtuality, essentially then the task is that of assessing the information processing consequences of the different dimensions of virtuality. More specifically, we need to locate the elements of virtuality in the information processing model. This implies building an understanding of how elements of virtuality affect the coordination load and capacity of an organization's existing design. To build this understanding we need to assess the *virtual information processing load:* that is, how each element may affect the amount of uncertainty and equivocality to which an organization is exposed, and the *virtual information processing capacity* being the ability of the organization to reduce uncertainty and resolve equivocality, respectively.

In the review of virtuality we identified two overall dimensions: locational and relational differentiation that represent virtuality well. As we briefly alluded to, they seem to have different impacts on organizations' coordination patterns. While relational differentiation had a direct effect on the ease of communication, the effect of locational differentiation was mediated by restrictions in the use of certain types of communication media.

From the review of the information processing perspective on organization design, and relative to Fig. 6.2, it seems appealing to locate the effects of relational and locational differentiation in the information processing model as changes in the interdepartmental relations and as changes in the communication media available to support decision-making and coordination processes, respectively. Accordingly, and relative to Daft and Lengelś information processing model, the two main effects of virtuality are argued to be (a) an increase in the information processing load caused by increased differentiation among departments and (b) a decrease in the information processing capacity caused by constraint put on the use of particular communication media. We will now turn to elaborate on these two dimensions of virtualness and their impact on information processing load and capacity.

6.3.1 Locational Differentiation and Its Information Processing Impacts

In the literature review, we defined locational differentiation in terms of the disabling effects of physical and temporal dispersion on the use of certain communication media among organizational units. Thus, relative to the information processing model presented by Daft and Lengel (1986), locational differentiation specifically addresses the effect of communication media on an organization's information processing capacity.

Being spatially differentiated even by small distances rapidly raises the cost and difficulty of communicating face-to-face. This fact particularly limits the variety and number of cues entailed in large parts of the communication activities among organizational units. Accordingly, the main effect of spatial differentiation is to constrain the use of face-to-face communication, which in turn decreases the ability of the organization to process rich information and hence reduce equivocality.

When spatial differentiation at the same time results in differences in time zones among organizational units, the cost and difficulty of using rich communication media increases even more. Additional to face-to-face communication, times zone differences also constrain the use of media such as phone- or video-meetings, which can now only take place at specific planned time periods. Further, use of e-mail and voicemail will now take a much more asynchronic character.

Differences in time zones mean that work in most instances is performed asynchronically by various units. Thereby the total number of hours in which communication may involve immediate feedback is reduced. As argued by Zack (1993), asynchronically organized teams often communicate through media that are lean, low in social presence, and low in interactivity.

Relative to media richness theory, decreasing the ability to provide immediate feedback, and decreasing the level of personalization, causes a reduction in the ability to process rich information and thereby reduces the capacity to resolve equivocality.

As a supplement to the original work by Daft and Lengel (1986) on media richness, work by Dennis and Valacich (1999) and Dennis et al. (2008) proposes three media characteristics to capture specific communicative capabilities of particularly new media. First, "parallelisms" refers to "the number of simultaneous conversations that can exist effectively." Second, "rehersability" refers to "the extent to which the media enables the sender to rehearse or fine tune the message before sending." And finally, "reprocessability" refers to "the extent to which a message can be reexamined or processed again within the context of the communication event" (Dennis and Valacich 1999).

Not only do these three characteristics add well to the characteristics describing media richness in its traditional form. Just as important in this context, the theoretical grounding of Dennis et al.'s theory of media synchonicity translates well into the information processing view proposed by Daft and Lengel (1986). In much a similar vein to Daft and Lengel (1986), Dennis and Valacich (1999) and Dennis et al. (2008) argue that communication processes have two main purposes: conveyance of information and convergence of meaning. Compared to Daft and Lengel (1986), conveyance can here been seen as a communication process intending to reduce uncertainty, while convergence of meaning can be seen as a process of reducing equivocality.

We find that particularly the effects of rehersability and reprocessability are important in distinguishing between the information processing capabilities on synchronous versus asynchronous communication media. As argued by Dennis and Valacich (1999), the use of asynchronous communication increases both rehersability and reprocessability, which in turn may have a positive effect on the ability of an organization to handle equivocal information. As such, when more time is used in between messages to interpret the incoming message and prepare for an outgoing message potential misunderstanding and conflicts are likely reduced. In that way asynchronous media may have a higher capacity for reducing equivocality, yet have a poorer fit for reducing uncertainty.

When spatial differentiation is accompanied by temporal differentiation, the constraint on the use of media does not have a straight forward effect on the capacity to resolve equivocality and reduce uncertainty, respectively.

The effect of spatial differentiation is to put constraints on the use of rich media thereby reducing the capacity for resolving equivocality while at the same time enhancing the capacity for processing lean information and handling uncertainty. The main effect of temporal differentiation is to constrain the use of synchronic communication media. The constraint on the use synchronic media, apart from reducing media richness, also leaves room for a higher level of rehersability and reproducibility, which in turn has a positive effect on the ability of the organization to process equivocal information. At the same time, temporal differentiation increases the time it takes to convey information and hence reduces the speed

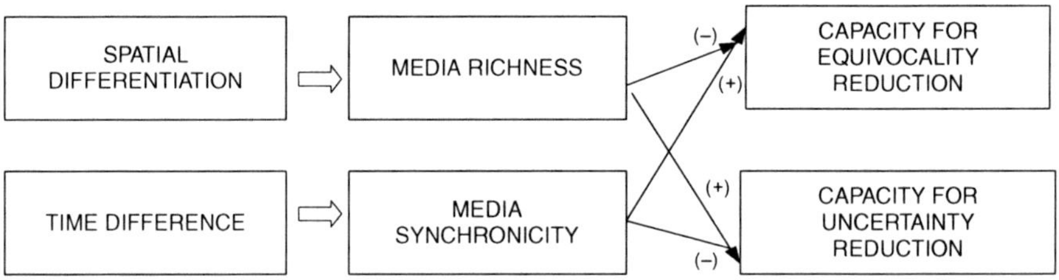

Fig. 6.3 Locational differentiation and its consequences for information processing

of reducing uncertainty among organizational units. Figure 6.3 sums up these information processing effects of spatial and temporal differentiation, respectively.

6.3.2 Relational Differentiation and Its Information Processing Impacts

In the literature review, we identified four kinds of differentiation (functional, cultural, inter-organizational, and temporary differentiation) that each have a significant impact on the character of the relations among virtually organized units. Each of the four elements of relational differentiation defines a potential for differences in cognitive orientations among organizational units and thus a potential for misunderstandings and conflicts among units.

In information processing terms, the presence of multiple and diverse cognitive orientations relates to the concept of equivocality which defines the "*multiplicity of meaning conveyed by information about organizational activities*" (p. 211, Daft and Macintosh 1981). As opposed to uncertainty, which leads to the search for and acquisition of external information, equivocality leads to the exchange of subjective understandings of information already available (Daft et al. 1987). Thus, the concern of equivocality is more related to the quality as opposed to the amount of information.

The convergence of subjective understandings is a process that often involves frequent feedback, use of multiple cues, high language variety, and messages infused with personal feelings and emotions. Hence, the ability of organizations to resolve equivocality has often been related to the richness of communication media available (Daft et al. 1987).

In that way, increasing functional, cultural, temporal and inter-organizational differentiation may necessitate the use of more rich communication media. Importantly, however, neither functional, cultural, temporal, nor inter-organizational differentiation directly affects the possibility of using any type of communication media.

The direct effect of relational differentiation is to increase the need to resolve equivocality among organizational units. Accordingly, and relative to the information processing model by Daft and Lengel (Fig. 6.2), relational differentiation increases the information processing load on the organization (Fig. 6.4).

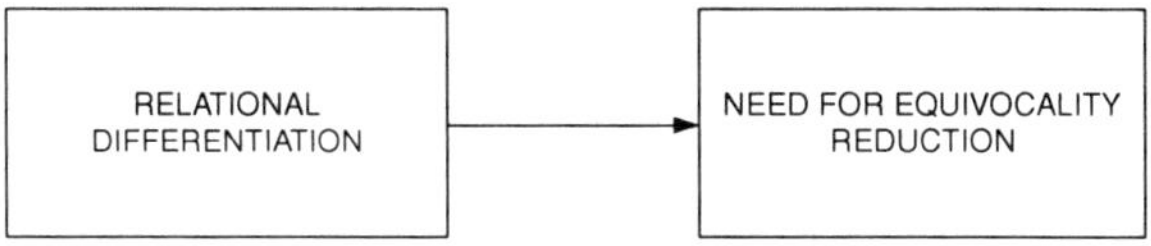

Fig. 6.4 Relational differentiation and its consequences for information processing

6.4 Discussion

We have now established that virtual characteristics conceptually divide into two main dimensions: locational and relational differentiation. While there may be a positive association between the two dimensions, each have a separate effect on the information processing structures of virtual organizing.

Locational differentiation includes synchronical and spatial differentiation and affects the information processing capacity of an organization by restricting the use of certain kinds of communication media.

Relational differentiation includes factors describing different kinds of cognitive differentiation among organizational units. These include functional, cultural, inter-organizational, and temporal differentiation. Relational differentiation does not by itself restrict the option of using certain communication media. However, this dimension increases the difficulty of communication as communicating partners become increasingly differentiated in their interests, values, and beliefs.

From that perspective, relational differentiation may at first be seen as a mediator to the effect of not being able to use certain communication media. The higher the relational differentiation on any of the four parameters, the more negatively the constraints on communication media affects communication effectiveness. However, from an information processing perspective, the effect of relational differentiation on organization design is more correctly viewed as a direct effect on the information processing load, not as a moderating effect on the effect of communication media on the information processing capacity of an organization.

From an information processing perspective, relational differentiation increases the information processing load on the organization by increasing the differences in goals, values, priorities, and cognitive elements. Thereby the level of ambiguity in the interdepartmental communication increases and leaves the organization with a higher requirement to reduce equivocality.

This effect of relational differentiation on the information processing requirement is to be theoretically distinguished from the effect that locational differentiation (mediated by constraints on the use of communication technology) has on the information processing capacity. As such, the effect of each of the two dimensions is to be located on each their side of the fit equation as illustrated in Fig. 6.5.

This conceptualization of virtuality clearly is to be distinguished from those who see both spatial differentiation and the use communication technology as defining dimensions of virtuality. As illustrated in Fig. 6.5, spatial (and synchronical) differentiation and communication technology are to be found on the same dimension,

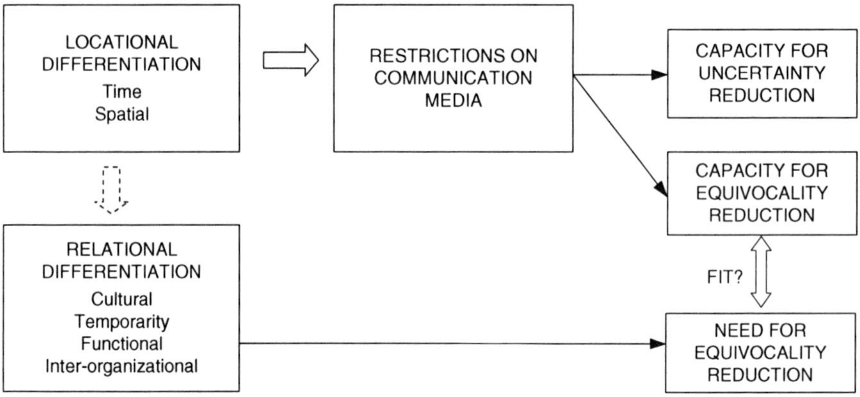

Fig. 6.5 Virtuality and its information processing consequences

the constraints in the use of communication technology being a mediating effect of spatial differentiation on the capacity of the organization to process information.

Second, the two-dimensional conceptualization of virtuality presented in this chapter is to be distinguished from conceptualizing virtuality merely in terms of differentiating elements. Indeed, while spatial and synchronical differentiation do work to separate organizational departments, this effect occurs through the down pointing arrow in Fig. 6.5 and hence works through its effect on relational differentiation on the information processing requirements. Opposingly, the effect of spatial and synchronical differentiation – through putting constraints on the use of particular communication media – that is, on the information processing capacity of the organization is one of reducing the ability of the organization to integrate across departments.

From this perspective, we can now see the challenges of virtual organization design in terms of the traditional organization design mechanisms of differentiation and integration. Furthermore, using the information processing perspective on fit and misfit makes it possible to integrate theory of virtual organizing into extant organization design theory.

Seemingly, virtuality according to Fig. 6.5 imposes an information processing dilemma for the organization in the sense that virtuality along the two dimensions (locational and relational differentiation) at the same time increases the information processing load and reduces the information processing capacity.

It is possible that the increase in functional specialization, which presumably is a lead motivator of organizing virtually, is accompanied by an increase in the local information processing capacity of separate organizational units, and that this increase in the local information processing capacity more or less compensates for the increased coordination load that is imposed among units.

However, from an organization design perspective, we still need to consider the potential misfit situation created for the organization as a whole by the simultaneous increase in information processing requirements and decrease in information processing capacity.

6.5 Conclusion

In this chapter, we have argued for the importance of understanding the nature and effects of the dimensions underlying virtual organizations, rather than simply focusing on how these dimensions are different from co-located organizations.

Toward this end, we have argued how previous characteristics of virtual organizations essentially can be captured by two overall dimensions of virtuality: locational and relational differentiation. Because these two dimensions relate to different coordination issues, the proposal of these two dimensions reveals some of the inherent conflict in much of the literature on virtuality.

In order to elaborate our understanding of how these two dimensions affect an organization's coordination needs, we theoretically anchor these dimensions to organization design and information processing theory. Through this anchoring, we demonstrate how the two dimensions essentially relate to different information processing outcomes: While locational differentiation reduces the organization's information processing capacity relational differentiation increases the organization's information processing need.

This imposes an interesting dilemma of coordinating virtual organizations which previous studies have not revealed. Increasing levels of virtuality directly opposes the two seminal strategies suggested for handling increasing levels of uncertainty in the environment: (1) increase information processing capacity and (2) reduce the need for information processing (Galbraith 1973). From an organization design perspective, such imbalances obviously need to be considered, by their own implications as well as by the implications they will have on other design parameters.

Thus overall, we hope to serve toward a better understanding of the coordination effects of virtual organizations.

6.5.1 Implications for Practice

Anticipating the future, virtuality seems likely to enter organizations in an even greater variety of ways and to an even greater extent than what we have seen previously. For example, merged organizations and strategic alliance partners may by history have several locations. While the motivation behind merger and alliance strategies may involve aspects of power and control (Pfeffer and Salancik 1978), the ability to put through any such strategy relies on the ability of the organization to implement organization structures in which decision-making and coordination efficiently and effectively supports the operational intentions behind the strategy. Accordingly, when considering and effectuating strategies that change the degree of virtuality within an organization, managers are in need of a framework that carefully considers the consequences of this change for the decision-making and coordination structures of the organization.

We believe that our way of conceptualizing virtuality has great managerial appeal in the sense that it not only reduces the complexity of the many descriptions into just two dimensions, but more importantly by integrating these two dimensions into Information Processing theory, it allows for a more tangible comprehension of the potential coordination problems associated with these. This is made applicable by describing the particular information processing consequences of the two dimensions and how they should potentially match. Moreover, by acknowledging that virtuality should be referred to as a degree, it is clear that these results have implications for managers of large multinationals as well as for managers of smaller teams – virtuality as a degree would apply for all of these.

6.5.2 Implications for Organizational Design

The implications for organizational design are related to a new understanding of how to conceptualize virtuality. This understanding is tied closely to the understanding of how different types of virtuality impacts an organization's coordination processes. Thereby, virtuality is not treated as an abstract phenomenon of a rare, prototypical design, but instead in information processing and coordination implication terms. Furthermore, by illuminating the inherent contradictions of the information processing implications of these two dimensions, we believe organization design theory serves to advance the understanding of virtual organizations.

We need to consider the misfit between coordination load and capacity created by relational and location differentiation respectively, not only in terms of its causes of but also in terms of what other design parameters are affected by this misfit and what other design parameters may potentially be adjusted to fit information processing requirements and capacity. As an example, it is likely, that a reduction in the interdependence among departments by for example introducing a more modular product architecture may reduce the information processing requirement. Similarly, a change in the use of coordination structures such as going from direct supervision to the use of integrators may increase the information processing capacity of the organization, hence again creating a fit situation.

The information processing model of virtual organization developed in this chapter can be restated as testable hypotheses for empirical research in the field. Given the embeddedness of the model in extant organization design theory, it is possible to extend the model to include organization design parameters others than those included in the two dimension of virtuality. For example, resent research has provided conflicting evidence on the role of hierarchical structures in virtual organizations (Hinds and McGrath 2006).

We believe that empirical studies as well as simulation experiments may give insights into the strength of the two opposing forces characterizing information processing in virtual organizing, and that this insight can help resolve current controversies on the effectiveness and efficiency of various coordination and control structures for the virtual organization.

References

Ahuja MJ, Carley KM (1999) Network structure in virtual organizations. Organization Science 10 (6): 741–757.

Alavi M (1993) Making CASE an organizational reality: Strategies and new capabilities needed. Information Systems Management 10 (Spring): 15–20.

Allen TJ (1977) Managing the Flow of Technology: Technology Transfer and the Dissemination of Technological Information Within the R&D Organization. Cambridge MA: MIT Press.

Armstrong DJ, Cole P (2002) Managing distances and differences in geographically distributed work groups. In: Hinds P, Kiesler SB (eds) Distributed Work. Cambridge, MA: MIT Press.

Arrow KJ (1974) The Limits of Organization. New York: W.W. Norton & Company.

Brewer MB (1996) In-group favoritism: The subtle side of intergroup discrimination. In: Messick DM, Tenbrunsel A (eds), Codes of Conduct: Behavioral Research and Business Ethics. New York: Russell Sage Found, pp 160–71.

Burton RM, Obel B (2004) Strategic Organizational Diagnosis and Design – The Dynamics of Fit. Dordrecht: Kluwer Academic Publishers.

Burton RM, DeSanctis G, Obel B (2006) Organizational Design: A Step-by-Step Approach. Cambridge, MA: Cambridge University Press.

Chesbrough HW, Teece DJ (1996) When is virtual virtuous? Harvard Business Review 74: 65–73.

Chudoba KM, Wynn E, Lu M, Watson-Manheim MB (2005) How virtual are we? Measuring virtuality and understanding its impact in a global organization. Information Systems Journal 15 (4): 279–306.

Cohen WM, Levinthal DA (1990) Absorptive capacity: A new perspective on learning organizations. Organization Science 7: 119–135.

Cramton CD (2001) The mutual knowledge problem and its consequences for dispersed collaboration. Organization Science 12 (3): 346–371, May/June.

Daft RL, Lengel RH (1986) Organizational information requirements, media richness and structural determinants. Management Science 32: 554–571.

Daft RL, Lengel RH, Trevino LK (1987) message equivocality, media selection and manager performance: Implications for information systems. MIS Quarterly 11: 355–366.

Daft RL, Macintosh NB (1981) A tentative exploration into the amount and equivocality of information processing in organizational work units. Administrative Science Quarterly 26 (2): 207–224.

Daft RL, Wiginton J (1979) Language and organization. Academy of Management Review 4 (2): 179–191.

Davidow WH, Malone MS (1992) The Virtual Corporation. New York: Edward Burlingame Books/Harper Business Press.

Dennis AR, Valacich JS (1999) Rethinking media richness: Toward a theory of media synchronicity. Proceedings of the 32nd Hawaii International Conference on System Sciences. Los Alamitos, CA: IEEE Computer Society Press.

Dennis Alan R, Fuller Robert M, Valacich J (2008) Media, tasks and communication processes: A theory of media synchronicity. MIS Quarterly 32 (3): 575–600, Sept.

DeSanctis G, Monge P (1999) Introduction to the special issue: Communication processes for virtual organizations. Organization Science 10 (6): 693–703.

Finholt T, Sproull L, Kiesler S (1990) Communication and performance in Ad Hoc Task Groups. In: Galegher J, Kraut RE, Egido C (eds), Intellectual Teamwork – Social and Technological Foundations of Cooperative Work. Hillsdale, NJ: Erlbaum.

Frey SC, Schlosser MM (1993) ABB and Ford: Creating Value through Cooperation. Sloan Management Review 35: 65–72.
Fulk J, DeSanctis G (1995) Electronic communication and changing organizational forms. Organization Science 6 (4): 337–349.
Galbraith J (1973) Designing Complex Organizations. Reading MA: Addison-Wesley.
Galbraith J (1974) Organization design: An information processing view. Interfaces 4 (3): 28–36.
Griffith TL, Neale MA (2001) Information processing in traditional, hybrid, and virtual teams: From nascent knowledge to transactive memory. Research in Organizational Behavior 23: 379.
Griffith TL, Sawyer JE, Neale MA (2003) Virtualness and knowledge in teams: Managing the love triangle of organizations, individuals, and information technology. MIS Quarterly 27(2): 265–287.
Grinter R, Herbsleb J, Perry D (1999) The geography of coordination: Dealing with distance in R&D work. Proceedings of GROUP '99. New York: ACM Press.
Guzzo RA, Dickson MW (1996) Teams in organizations: Recent research on performance and effectiveness. Annual Review of Psychology 47 (1): 307.
Handy C (1995) Trust and the virtual organization. Harvard Business Review 73: 40–50.
Herbsleb J, Mockus A, Finholt T, Grinter R (2000) Distance, dependencies, and delay in a global collaboration. Proceedings of CSCW 2000. New York: ACM Press.
Hinds PJ, McGrath C (2006) Structures that Work: Structure and Coordinated Ease in Geographically Distributed Teams. Group and Organization Interfaces, 343–352.
Horii T, Jin Y, Levitt RE (2005) Modelling and analyzing cultural influences on project team performance. Computational and Mathematical Organization Theory 10: 305–321.
Kirkman BL, Mathieu JE (2005) The dimensions and antecedents of team virtuality. Journal of Management 31 (5): 700–718.
Kramer R (1999) Trust and distrust in organizations: Emerging perspectives, enduring questions. Annual Review of Psychology 50: 569–598.
Kraut RE, Fussell SR, Brennan SE, Siegel J (2002) Understanding effects of proximity on collaboration: Implications for technologies to support remote collaborative work. In: Hinds P, Kiesler S (eds), Distributed Work. Cambridge, MA: MIT Press, pp 137–162.
Kumar J (2006) Working as a Designer in a Global Team. Interactions 13: 25–26, March, April 25–27.
Lawrence PR, Lorsch JW (1967) Organization and Environment. Boston, MA: Harvard Business Press.
Lipnack J, Stamps J (1997) Virtual Teams: Researching Across Space, Time and Organizations with Technology. New York: John Wiley and Sons.
Malone TW, Crowston K (1994) The interdisciplinary study of coordination. ACM Computing Surveys 26 (1): 87–120.
Martins LL, Gilson LL, Maynard TM (2004) Virtual teams: What do we know and where do we go from here? Journal of Management 30 (6): 805–835.
Monge P, DeSanctis G (1999) Introduction to the special issue: Communication processes for virtual organization. Organization Science 10 (6): 693–703, Nov.–Dec.
Mowshowitz A (1997) Virtual organization. Communications of the ACM 40 (9): 30–37
Mowshowitz A (1994) Virtual organization: A vision of management in the information age. The Information Society 10: 267–288.
Nohria N, Berkley JD (1994) The virtual organization: Bureaucracy, technology, and the implosion of control. In: Heckscher C, Donnelon A (eds), The Post-Bureaucratic Organization: New Perspectives on Organizational Change. Thouand Oaks, CA: Sage, pp 108–128.
O'Leary MB, Cummings JN (2007) The spatial, temporal, and configurational characteristics of geographical dispersion in teams. MIS Quarterly 31 (3): 433–452.
Perrow C (1967) A framework for the competitive analysis of organizations. American Sociological Review 32: 194–208.
Pfeffer J, Salancik GR (1978) The External Control of Organizations: A Resource Dependence Perspective. New York: Harper & Row.

Reagans R, McEvily B (2003) Network structure and knowledge transfer: The effects of cohesion and range. Administrative Science Quarterly 48 (2): 240–267.
Rogers E, Bhowmik D (1970) Homophily-heterophily: Relational concepts for communication research. Public Opinion Quarterly 34: 523–538.
Simon HA (1991) Bounded rationality and organizational learning. Organization Science 2 (1): 125–134.
Thompson JD (1967) Organizations in Action. New York: McGraw-Hill.
Tushman ML, Nadler D (1978) Information processing as an integrating concept in organizational design. Academy of Management Review 3: 613–624.
Venkatesh V, Johnson P (2002) Telecommuting technology implementations: A within-and between-subjects longitudinal field study. Personnel Psychology 55: 661–687.
Watson-Manheim MB, Chudoba K, Crowston K (2002) Discontinuities and continuities: A new way to understand virtual work. Information, Technology and People 15: 191–209.
Zack MH (1993) Interactivity and communication mode choice in ongoing management groups. Information Systems Research 4(3): 207–239.

Part III
Fit and Performance

Chapter 7
Learning-Before-Doing and Learning-in-Action: Bridging the Gap Between Innovation Adoption, Implementation, and Performance

Eitan Naveh, Ofer Meilich, and Alfred Marcus

Abstract Implementation links purpose to outcome. Our model of implementation effectiveness centers on learning – learning-before-doing (preparation) and learning-in-action (adaptation and change catalysis). We explain both the degree of implementation and its impact on various measures of performance (subjective and objective) and test our proposed model on a large, multi-industry sample in the context of implementing the ISO 9000 quality standard. We find that learning-before-doing, an important means for bridging the adoption–implementation gap, is a necessary but not sufficient condition for realizing the benefits of a planned change. To fully bridge the implementation–performance gap, both aspects of learning-in-action – adaptation and change catalysis – must accompany implementation.

Keywords Implementation · Learning · ISO 9000

7.1 Introduction

Implementation is the series of actions that organizations take to bring about intended change (Nutt 1986: 230). Organizational research has extensively documented the importance of implementation (e.g., Bardach 1977; Bossidy and Charan 2002; Nutt 1989; Pfeffer and Sutton 1999; Reger, et al., 1994). Just as planned change is ubiquitous in organizational life so too is implementation. It is relevant to a wide array of change processes (Fidler and Johnson 1984; Nutt 1986), from upper level strategy making to lower level decision-making. It has bearing on governmental initiatives and organizations diffusing technological and administrative innovations, and plays a central role in reorganization and organizational regeneration.[1]

E. Naveh (✉)
Faculty of Industrial Engineering and Management, Technion – Israel Institute of Technology, Haifa 32000, Israel
e-mail: naveh@ie.technion.ac.il

[1] From hereon, the terms "change" and "innovation" are used interchangeably, to mean potentially beneficial alterations in the way an organization functions.

A. Bøllingtoft et al. (eds.), *New Approaches to Organization Design*, Information and Organization Design Series 8, DOI 10.1007/978-1-4419-0627-4_7,

Organizational research has shown that the intention to make a change does not necessarily mean that the change will come about. There must be follow up if the intended change is to become part of the organization's routines and everyday activities. These follow-up actions are often not forthcoming. Implementation failure is seen as one of the main causes of organizations' inability to realize the intended benefits of changes they make (Klein, et al., 2001). Aiman-Smith and Green (2002) found that the estimated failure rate for implementing new technologies was 47%. Improvements in organizational performance, according to Pfeffer and Sutton (1999), hinge not on finding new knowledge but on implementing what is known.

Repening (2002: 109) asserted that "existing theory offers little on the question of why potentially useful innovations often fail to find a permanent home in the organizations that try to implement them." Early research concentrated on the *adoption–implementation gap*, asking why innovations adopted are not thoroughly and completely institutionalized (Nutt 1986, 1989). Such studies examined implementation as a dichotomous dependent variable and assumed that full and complete implementation would lead to better organizational outcomes (Lewis and Seibold 1993). Later researchers started to question this assumption. They dealt explicitly with the *implementation–performance gap*. For instance, Abrahamson's research on fads and fashion (Abrahamson 1991, 1996, 1997) documented the lack of improvement following adoption of administrative innovations. More recently, researchers have begun to employ measures of the *extent of implementation* as opposed to dichotomous measures of either complete or nonexistent implementation (Klein, et al., 2001; Douglas and Judge 2001).

This chapter continues in this promising vein. We follow the footsteps of researchers who are not interested in the issue of extent of implementation per se nor in the performance of a planned change if fully carried out. Rather, our aim is to investigate in what ways extent of implementation has an effect on performance. We incorporate measures of the degree of implementation as supposed to dichotomous measures, something very few past researchers have attempted (Douglas and Judge 2001). We explain both the degree of implementation and its impact of various measures of performance (subjective and objective) and test our proposed model on a large, multi-industry sample.

Our view of implementation is a process one, as opposed to a contextual or structural contingencies approach. The model of implementation effectiveness, which we construct and test, centers on learning – learning-before-doing (preparation) and learning-in-action (adaptation and change catalysis). We find that learning-before-doing, an important means for bridging the adoption–implementation gap, is a necessary but not sufficient condition for realizing the benefits of a planned change. To fully bridge the implementation–performance gap, both aspects of learning-in-action are needed.

This chapter proceeds as follows. In the next section, we develop our model of implementation. The following section depicts the methods including how we collected data and the measures. Then, we present the results. The final section discusses limitations, opportunities for further research, and implications for practitioners.

7.2 Implementation: A Conceptual Model

Our model of implementation (see Fig. 7.1) starts with an adoption decision. The gap between adoption and successful implementation is bridged by preparation (i.e., learning-before-doing) and the gap between implementation and performance is bridged by adaptation and catalysis – two types of learning-in-action. These two types of learning-in-action need to occur concurrently with implementation to realize the potential gains encapsulated in a change. In the rest of this section, we explain this model and the hypotheses we derive from it.

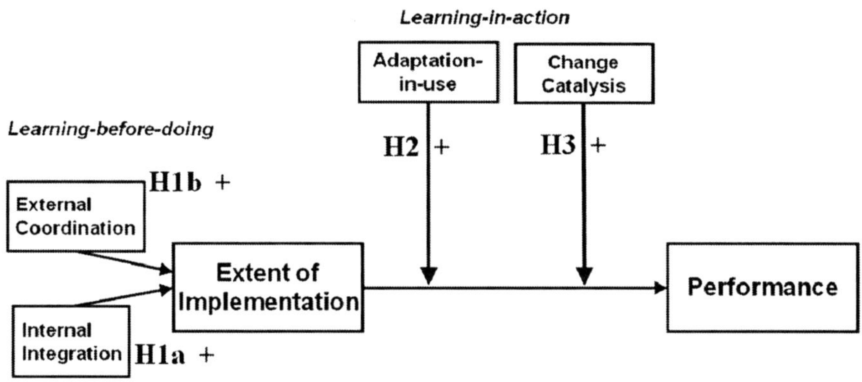

Fig. 7.1 An implementation model

7.2.1 The Adoption–Implementation Gap

Successful implementation requires that an innovation will become an integral part of an organization's daily practice (e.g., Beyer and Trice 1978; Edmondson et al. 2001; Fidler and Johnson 1984; Rogers 1995).[2] Preparation is an indication of an organization's intention to implement. Such activities entail what Pisano (1996) calls learning-before-doing, in the context of technological innovation. Pisano (1996: 1102) defines this term as the range of problem-solving activities that constitute the "attempts to predict process performance and to identify potential problems before the process is transferred to the factory." In essence, learning-before-doing, in the context of an administrative innovation or organizational change, constitutes a process of dynamic preparation. Preparation and learning-before-doing will be used from here on interchangeably.

[2]As Klein and Sorra (1996: 1057) comment: "Implementation is the transition period during which targeted organizational members ideally become increasingly skillful, consistent, and committed in their use of an innovation. Implementation is the critical gateway between the decision to adopt the innovation and the routine use of the innovation within an organization."

Learning before doing is expected to significantly contribute to successful implementation for obvious reasons – the greater the investment in making sure that the change is commensurate with internal and external requirements, the better are the likely results (Nord and Tucker 1987: 313–315). On the other hand, when the adoption of a change is done primarily for show, preparation will be superficial and the actual use of such a change is expected to be limited and shallow (Staw and Epstein 2000). Preparation is expected to enhance implementation via two routes. First, preparation reduces uncertainty because it provides a structured approach for decision making (Miller and Cardinal 1994). Second, the explicit attention given to thinking about the change before implementing it enhances coordination, commitment, and control within the organization (Steiner 1979).

Mirroring these two routes, preparation is conceived here as having two broad aspects – internal integration and external coordination (Nord and Tucker 1987). Internal integration means designing and developing systems that fit with a company's existing procedures (Meyer and Goes 1988; Brunsson, et al., 2000). An organization's old practices represent its past knowledge, which was used to meet prior needs (Matusik and Hill 1998), but links have to be established between these old practices and the new ones (March, et al., 2000). The second aspect of preparation is external coordination. External coordination involves an attempt to harmonize the change with the requirements and expectations of external stakeholders. With supply chains increasingly critical to an organization's success, external coordination takes on growing importance (Eisenhardt and Tabrizi 1995; Lengnick-Hall 1996).

During the design and development phase, the more an organization invests in integration of the new practices with the policies and procedures that are already have in place, and the further it attempts to coordinate the new practices with external parties, the more encompassing will be the implementation of the change. Therefore, we propose that

H1a: Internal integration increases the extent of implementation.
H1b: External coordination increases the extent of implementation.

7.2.2 The Implementation–Performance Gap

Organizational learning, we posit, is the key to realizing gains from innovation. When implementation is conceived as aiming to achieve a static "end state," the importance of learning is downplayed (Lewis and Seibold 1993). We argue that implementation by itself is a necessary but not sufficient condition to gain from a potentially beneficial change. Implementation needs to be complemented by ongoing reflection, re-adaptation, and innovation (Crossan et al. 1999; Sitkin et al. 1994; Nord and Tucker 1987).[3] The type of learning, which is most relevant to

[3]Orlikowski (2002: 250) takes an even stronger view that stresses that knowledge and practice reciprocally constitute each other.

implementation, is *learning-in-action.*[4] Learning-in-action is defined as "the cyclical interplay of thinking and doing" (Carroll, et al., 2003: 575). It is the most effective form of learning because it minimizes the discrepancy between knowing and doing (Pfeffer and Sutton 1999).

When learning-in-action occurs, everyday practices are the "template" (Moorman and Miner 1998a, 1998b) around which experimentation and additional learning takes place (Mirvis 1998; Weick 1998). Such learning happens in communities-of-practice in context through collaboration and the social construction of meaning (Brown and Duguid 1991; Lave and Wenger 1991; Wenger, McDermott and Snyder 2002). This type of learning highlights the existence and desirability of concurrent action and reflection (Schön 1983).

We distinguish between two types of learning-in-action: *adaptation-in-use* and *change catalysis*. In the context of implementation, these types are comparable to Argyris and Schön's (1978) single-loop and double-loop learning. Adaptation-in-use is defined as adaptation of the innovation once implementation activities have started. Adaptation-in-use aims to find a better fit by adapting both the innovation and the organization to the specific context in which the change is implemented. Change catalysis is defined by a situation where the currently implemented innovation itself is being used for further change and innovation.[5] Change catalysis is a form of deep learning in which as implementation happens, it sets in motion a process in which basic assumptions, models, norms, policies, and objectives are challenged (Carroll et al. 2003).

The two types of learning-in-action also mirror March's (1991) distinction between exploration and exploitation. Adaptation-in-use involves proximal search, while change catalysis embodies far-reaching search. Similar contrasts are exhibited between incremental learning and step function learning, the latter "involves fundamental changes to core or integrative knowledge" (Helfat and Raubitschek 2000: 967), evolutionary and revolutionary change (Tushman and O'Reilley 1996), operative and strategic learning (Thomas et al. 2001), and adaptive and generative learning (Chelariu et al. 2001). We employ the terms adaptation-in-use and change catalysis to indicate the organizational actions that are specific to the context of implementation.

7.2.2.1 Adaptation-in-Use, Implementation, and Performance

As comprehensive as preparation activities may be, not all conditions and contexts of use can be foreseen. Once an innovation is put to use, the need to adapt the innovation and the organization arises (Rogers 1995; Szulanski and Cappetta 2003). As Edmondson et al. (2001: 686) note, "the adopting organization must go through

---

[4]We avoid using the term "learning-by-doing" even though it is close to "learning-in-action" because learning-by-doing is identified with learning curve effects (Argote, 1999).

[5]This concept expands on Greve and Taylor's (2000) idea of technological innovation as being a catalyst for additional organizational changes.

a learning process, making cognitive, interpersonal, and organizational adjustments that allow new routines to become ongoing practice." The new practices need to be embedded in their specific context (Fidler and Johnson 1984). The knowledge embedded in the abstract model to be implemented must be integrated with existing practices and with their underlying knowledge bases.

Because organizations are rather complex entities, not all contingencies may be foreseen; so adaptation-in-use must take place. The unforeseen changes are related to a plethora of organizational phenomena. They are related to organizational members' tasks and roles, work arrangements, the pattern of interdependence, communication networks, authority relations, and distribution of status and expertise (Lewis and Seibold 1993). Once employed, the physical manifestations of the innovation itself and the nature of use of the innovation necessitate re-adaptation. These cycles of change-reflection-adaptation entail incremental changes that do not challenge the working assumptions behind the innovation itself. Therefore, adaptation-in-use activities fit Argyris and Schön's (1978) notion of single-loop learning as they strive to readjust the organization and the innovation to each other. Adaptation-in-use also suggest local search for solutions to exploit the innovation. "Exploitation," writes March, "includes such things as refinement, choice, production, efficiency, selection, implementation, execution" (1991: 71).

We contend that, to realize the potential gain of an innovation, implementation must be accompanied by adaptation-in-use. Such ongoing adaptation should improve the realization of the benefits embedded in the innovation, and should lead to enhanced organizational performance (e.g., Marcus 1988). Therefore, we hypothesize that

> H2: Adaptation-in-use moderates the relationship between implementation and subsequent performance; the greater the extent of adaptation-in-use, the more positive is implementation's effect on performance.

7.2.2.2 Change Catalysis, Implementation, and Performance

While the need for adaptation of the innovation and the organization during the implementation phase is well-documented, the use of the innovation itself as a facilitator of further change has received much less attention. Change catalysis involves using the currently implemented change as a springboard for rethinking the way the organization does business, an opportunity to innovate, and a starting point for the introduction of new and more advanced practices. Such explorative search entails what Carroll et al. (2003: 582) call deep learning – learning that involves "skillful inquiry and facility to gain insights, challenge assumptions, surface existing frames, and create comprehensive models." Similarly, Thomas et al. (2001: 332) refer to it as strategic learning, a type of learning that "leverages the organization's ability to generate, store, and transport rich de-embedded knowledge across multiple levels for the purpose of enhancing firm performance." These researchers see a direct relation between change catalysis and performance. We, on the other hand, maintain that the interaction between implementation and change catalysis is a significant source of organizational performance. This improvement happens in two ways.

First, change catalysis activities can make implementation more rewarding. Nord and Tucker (1987: 305, 311) found that implementation of radical and incremental innovations in financial institutions allowed firms to discover organizational deficiencies and change them in ways that included redefining these firms' fundamental policies. Similarly, Sitkin et al. (1994) suggested that implementation success depends on more than everyday use. Independent thinking and initiative are needed, for without them (Fidler and Johnson 1984: 704), there would be "routine" and "mechanical" implementation.

An explanation for the effects of change catalysis is the overcoming of the always-powerful inertial tendencies that lie deeply rooted in organizations. As previously defined, implementation turns change into a routine part of an organization's operations. Routinization tends to lock organizations into competency traps (March 1991), habitual routines (Gersick and Hackman 1990), and core rigidities (Leonard-Barton 1995). Engaging in activities that question the fundamental assumptions creates awareness and a state of mind that rejects inertia in an attempt to improve performance and serves as a mechanism to avoid getting stuck into the habitual (Pfeffer and Sutton 1999).

Implementation also increases the likelihood that exploratory search will yield beneficial results. Compared to exploitation, exploration activities carry a high potential return, but also greater risk (March 1991). Implementation activities can act as an organizational resource (Lewis and Seibold 1993: 333) to reduce the risk inherent in exploration. Implementation builds practice-based knowledge. Such knowledge can be more readily translated to enhanced performance, as compared to abstract knowledge (Pfeffer and Sutton 1999). Knowledge catalyzed by implementation is deeper and more profound. Just like in playing jazz, success in being innovative depends on the performer's competency in the elemental, normal, and routine use of the instrument (Barrett 1998). Similarly, implementation reveals to the performers (either developers or users) their own skills, the skills of their colleagues, the potential encapsulated in the current innovation, and the structure in which they are embedded (Argote 1999).

Implementation can improve communication and establish collective norms and goals, facilitating the sharing of ideas in communities-of-practice (Brown and Duguid 2001). According to Greve and Taylor (2000: 57), "Interpretations, meaning, valuations, and the relevance of future options shift when organizations act." Hence, the implementation of an innovation changes the organizational cognitions about the very appropriateness, feasibility, and necessity of taking action (Greve and Taylor 2000), and therefore can highlight more fruitful venues for additional and ongoing innovation. Therefore, we hypothesize that implementation and change catalysis synergistically interact to enhance organizational performance.

H3: Change catalysis moderates the relationship between implementation and subsequent performance; the greater the level of change catalysis, the more positive is implementation's effect on performance.[6]

7.3 Methods

7.3.1 The Context: ISO 9000 Quality Standard

The setting for our study revolves around the quality standard ISO 9000. New quality models are continually evolving and being refined over time, and such models are supposed to be a competitive factor in many industries (Cole 1999; Reger et al. 1994). ISO 9000 is a subset of a broader category of quality initiatives, which are often referred to as administrative innovations (Brunsson et al. 2000). It requires that organizations have verifiable routines and procedures in place for product design, manufacture, delivery, service, and support. They must strictly monitor the steps they take to complete a job (Cole 1999; Brunsson et al. 2000). The organizations are required to comply with procedures they establish for themselves. To guarantee compliance, third party auditors carry out site visits twice a year. ISO 9000 is the most commonly implemented quality standard in the world. By the year 2000, more than 100,000 organizations had been certified worldwide and over 100 countries had adopted ISO 9000 (Wayhan et al. 2002; see also Guler et al. 2002).

However, the results of its implementation are far from clear. While some analysts claim that the implementation of ISO 9000 has achieved real benefits, others say that it has accomplished little more than burdening its adopters with the cost of over-documentation and bureaucracy. For instance, Robin and Dennis (1994) found that ISO 9000 was effective in introducing statistical process control and improving business performance; Dun and Bradstreet's comprehensive survey of certified U.S. sites found that the average company benefited from ISO 9000 implementation (ISO 9000 survey 1996); Elmuti and Kathawala (1997) report that productivity, quality of product, and quality of work life improved due to certification; and Brown and Loughton (1998) found benefits of greater quality awareness, improved awareness of problems, and better product quality, but not improvements in productivity, costs, yield rates, staff motivation, and staff retention. On the other hand, many others found very little benefit from this standard. For example, Batchelor (1992) found that the benefits of certification where mainly procedural efficiency and error rate and not market share, staff motivation, or cost; Powell's (1995) international quality study concluded that certification had no significant effect on performance;

[6] H3 is operationalized in an identical manner to the following hypothesis – *Extent of implementation moderates the relationship between change catalysis and performance; the greater the extent of implementation, the more positive will be effect of change catalysis on performance.* Hence, the operationalization (as an interaction term's effect on performance) captures both arguments leading to H3.

Hunt's (1997) main conclusion was that ISO9000 did not guarantee product quality; and Terziovski et al. (1997) rejected the hypothesis that there was a significant relationship between ISO 9000 implementation and organizational performance.

We contend that some of the variability in the effects of implementing ISO 9000 may be due to unmeasured differences in preparation activities and to unmodeled interactive effects of implementation with both adaptation-in-use and change catalysis.

7.3.2 Data Collection

Our primary data collection vehicle was a widespread survey mailed to all ISO 9000-registered facilities in North America (Naveh et al. 1999). We designed the questionnaire in an iterative process. We started with an extensive literature review on ISO 9000. We then visited and interviewed eleven facilities of companies that had implemented ISO 9000.[7] We consulted with world leading experts on the ISO 9000 standard such as members of the TC 176 group (the international committee that is in charge of the standard), U.S. registrars, consultants, and professional quality managers. Additionally, we conducted a pilot study to which several dozens of respondents reviewed the questionnaire and provided comments.

The newsletter *Quality System Update* by McGraw Hill routinely lists all facilities in North America which have demonstrated through external audits that they have implemented ISO 9000. We sent a postcard to the ISO 9000 management representatives, which appeared in this list, asking these persons to participate in the study. They were given a code that gave them entry to an Internet site where a questionnaire could be completed (full confidentiality was promised). A month after the first postcard was mailed, we sent a second postcard to non-responders. The total number of ISO 9000 registrations in North America at the time the survey was 4233. Depending on an organization's decision, registration could be held at the organization level or at the facility level. In organizations with more than one registration, we sent a separate postcard to each registered facility.

In total 1,150 ISO 9000 facilities from 885 companies completed the survey, a response rate of more than 27%. Among surveys of quality managers, this response rate is typical: published empirical works by Flynn et al. (1994), Powell (1995), Douglas and Judge (2001) draw conclusions from surveys with similar response rates. Seventy-five percent of the responders were manufacturing facilities and the balance was service facilities. About a third of the companies (304 companies) that responded to the questionnaire had publicly available data in the financial dataset Compustat, which we used to augment the survey.

[7] Before the interviews, we constructed an interview guide that had open-ended questions about the company's experience with ISO 9000. During the site visit, we heard general presentations. We interviewed people from the functions of: auditing (both internal and external), quality management, manufacturing, engineering, software development, and documentation. Typically, the interviews lasted from 1 to 1.5 hours and were taped. We also took handwritten notes, and while on-site collected relevant documents. The interview team discussed its impressions with company representatives and had off-site debriefings.

Because of a concern that the survey respondents might have a different profile than non-respondents, we divided the respondents by industry sector (using 2-digit SIC codes), time since certification, and size (measured as number of employees), and compared the percentage responding in these groups with the actual registrants in these groups. The numbers were close enough for us to believe that those who responded came from similar groups as the population as a whole. In addition, we mailed questionnaires to a sample of 50 non-responding organizations, and followed up with telephone calls until we obtained responses from nearly all 50. Because the generalizability of the findings of this study rests in part on the extent to which the sample was free from non-response bias, a time trend extrapolation test (Armstrong and Overton 1977) was performed as a check on non-response bias. The assumption behind this test was that late respondents (those whose responses came from the second postcard and from the mailed questionnaire) would be very similar to non-respondents, given that they would have fallen into that category had a second set of postcards and questionnaires not been mailed. We used the multivariate general linear model (GLM) procedure to test the null hypothesis of no difference. The procedure simultaneously compared the three survey groups (responders after first and second postcard and for the mailed questionnaires) with respect to our study variables. This analysis indicated no difference (insignificant Wilks' Lambda). Although non-response bias cannot be ruled out, the results of the comparison testing of the second mailed postcard and mailed questionnaire increased our confidence in the representativeness of the sample. A similar comparison test for the responders' subgroup for which we had publicly available data (304 facilities) also showed similar proportions to the ISO 9000 certified population.

7.3.3 Measures

7.3.3.1 Independent Variables

Our five main independent variables (internal integration, external coordination, implementation, adaptation-in-use, and change catalysis) were all computed from the respondents' answers to 5-point Likert scale-type items.

Internal integration was calculated as the mean score of four questionnaire items related to the extent to which ISO 9000 was customized, tailored, and linked to existing practices during the design and development phase of the facility's ISO 9000 system. The questionnaire items we used are based on the work of Dierickx and Cool (1989). These items asked about the to extent which the design and development of the respondent's ISO 9000 system was: (1) Coordinated and led by employees who were trained and developed inside the organization; (2) Integrated with practices already in place; (3) Based on an analysis of internal processes and performance; and (4) Customized to the needs of your company. This variable's reliability (Cronbach alpha) was 0.92 for the full sample and 0.91 for the public companies sub-sample.

External coordination was operationalized as the mean score of three items related to the extent to which the design and development of the facility's ISO 9000 system was externally coordinated. The questionnaire items we used are based on Eisenhardt and Tabrizi (1995). These items asked about the extent to which the design and development of the respondent's ISO 9000 system was: (1) Coordinated with suppliers; (2) Coordinated with customers; and (3) Based on learning from other companies that already were registered? This variable's reliability was 0.91 (0.90 for the public companies).

Implementation was calculated as the mean score of five items, dealing with the extent to which ISO 9000-related procedures had become an integral part of the normal operations of the facility. The questionnaire items we used are based on Brunsson et al. (2000). These items referred to the extent to which: (1) The documents created for the purpose of ISO 9000 registration were used in daily practice; (2) Preparations for external audits were made at the last minute (reverse-scored); (3) The system was regularly ignored (reverse-scored); (4) The system was an unnecessary burden (reverse-scored); and (5) The system has become part of the regular routine? This variable's reliability was 0.93 (0.92 for the public companies).

Adaptation-in-use was calculated as the mean score of three items, related to the extent to which the original design of the ISO 9000 system was changed during implementation. The questionnaire items we used are based on Cooper and Zmud (1990). The items asked the following: (1) Have changes in your ISO 9000 system been made since registration? (2) Are the documents regularly updated? and (3) Has ISO 9000 changed daily practice? This variable's reliability was 0.94. (0.94 for the public companies).

Lastly, change catalysis was calculated as the mean score of five items, related to the extent to which the investments and use of the facility's ISO 9000 system served as a catalyst for change. The questionnaire items we used are based on Argote (1999). The first three items asked about the extent to which the responding facility's investment of time and resources in ISO 9000 was (1) A starting point for other more advances practices; (2) A catalyst for rethinking the way you do business; and (3) Understood as an opportunity to innovate. The two other items asked: (4) To what extent was design and development of your ISO 9000 system a springboard to introduce new practices? (5) To what extent has ISO 9000 led to the discovery of improvement opportunities? This variable's reliability was 0.92 (0.93 for the public companies).

Results of a confirmatory factor analysis indicate a good fit with the data ($\chi^2 = 421.52$, d.f. $= 316$). The goodness of fit index was 0.95, the comparative fit index 0.94, and the root-mean-square error of approximation was 0.053. Item loadings were as theorized and significant ($p < 0.01$). The confirmatory factor analysis results indicate that these five main dependent variables are empirically distinct from each other.

7.3.3.2 Dependent Variable – Performance

We employed four measures of performance. Operating performance was calculated as the mean score of five questionnaire items. Specifically, each facility representative was asked about the operational improvements that resulted from the implementation of ISO 9000 in the following five areas: defect rates, quality costs, productivity, on-time delivery, and customer satisfaction. All these items were on a 5-point Likert scale. Reliability of this measure was 0.89 (0.90 for the public companies).

As mentioned above, for about a third of the organizations we had Compustat data. Therefore, we were able to incorporate three company-level measures of the change in: cost of goods sold, sales, and gross profit margin (calculated as sales minus cost of goods sold divided by sales). Specifically, we calculated the percent change in cost of goods sold as the value 2 years after registration minus the value 2 years prior to their ISO 9000 registration,[8] divided by the average measure in the 2 years prior to registration. Sales change and gross profit margin change were calculated as their value 2 years after obtaining their ISO 9000 registration minus the same variable 2 years prior to registration, divided by the average measure in the 2 years prior to registration. This time frame of 2 years was based on Bagchi (1996) and Wayhan et al. (2002). We ran the analyses with 3, 4, and 5 years time frames and obtained similar results.

Because of the difference in levels of analysis on Compustat data (company as opposed to facility) we included only firms for which meaningful ISO 9000-related analyses could be performed. Specifically, we included public firms based on three categories. First, we included all publicly held firms that have only one facility which was ISO 9000 registered (249 firms). Second, we included firms with more than one ISO 9000 facility, where all facilities were registered and have been surveyed (35 firms). Last, we included firms that had more than one facility, one or more of these facilities was ISO 9000, and 95% of sales or higher originated from ISO 9000 facilities (20 firms). The second and third categories required averaging facility-level survey results. To ensure homogeneity of these multiple-facility items we calculated Rwg (within-group interrater agreement). This coefficient ranged between 0.81 and 0.91, above the threshold value of 0.70 (James et al. 1984).

7.3.3.3 Control Variables

Five control variables were included: (1) time since registration[9]; (2) number of employees – at the facility for the facility level of analyses and total company employment for the company-level analyses; (3) a dummy variable for industrial sector which was equal to one when a registered facility was classified as belonging

[8]In companies with more than one ISO 9000 registrations, we ran the analyses with measures corresponding to their first registration. For the 55 public companies, on the date we finalized the survey, the mean and standard deviation of the time passed since first facility registration were 43.0 and 15.6 months, respectively. Time since last facility registration was 41.8 months (14.9 standard deviation). We also ran analyses with measures relative to the most recent registration. Results were very similar to using the first registration time frame.

[9]This variable was incorporated only in the models predicting self-reported measures; it is constant for the Compustat performance measures because all time-relevant performance measures were obtained from Compustat.

to a service or a software company and otherwise equal zero; (4) external pressures, computed as the mean score of two 5-point Likert scale questionnaire items in which the respondent estimated quality demands from customers and from regulators ($\alpha = 0.87$; 0.88 for the public companies subset); and (5) technological uncertainty, computed as the mean score of three 5-point Likert scale questionnaire items in which the responded estimated, for their industry setting, the rate of product change, the rate of process change, and the amount of research and development ($\alpha = 0.90$; 0.90 for the public companies subset).

Several other control variables were included in the analyses but were not statistically significant (and hence are not reported here). These variables were: number of external audits to attain registration, ratio of registered to total number of facilities within a firm, employee education, product complexity, and marketplace competition. Three Compustat variables were also not statistically significant. These were R&D expenses, leverage, and capital Intensity.

Descriptive statistics and correlations of the full and public companies subset are presented in Table 7.1 below.

7.3.4 Analysis

We employed a series of OLS regressions, forming a path analysis model. In all regressions involving interaction terms the dependent variables were first normalized to obtain correct standardized coefficients, as recommended by Aiken and West (1991). Since ISO 9000 certification could be gained at the facility level (the questionnaire data in our study) while Compustat data is at the company level, a principal issue of concern was how to analyze and interpret data that resides at both within group levels and between group levels. As explained earlier (in the section describing the Compustat performance measures), the vast majority of companies included had a single facility which was ISO 9000-registered. When a company had more than one registered facility, we incorporated companies with 95% or more of sales coming from ISO 9000 facilities, and then we tested for interrater response homogeneity (Rwg).

In the models where all measures were based on questionnaire responses (self-report data) there might be a problem with common method bias. The Harman's single-factor test therefore was used to test for such bias (Harman 1967). The basic assumption is that if a substantial amount of common method variance is present, either a single factor will emerge from the factor analysis or one general factor will account for the majority of the covariance in the independent and dependent variables. We ran several separate exploratory factor analyses, combining items for the dependent variables with items for different sets of independent variables. In all of these analyses, we obtained more than one factor with eigenvalue greater than 1. In addition, items from the different constructs separated cleanly. No item from one construct had a loading greater than 0.5 on a factor associated with another construct. Therefore, we believe that common method bias in our analyses is minimal.

Table 7.1 Means, standard deviations, and correlations

Variable	Facilities (without compustat organizations)		Organizations (only compustat organizations)															
	Mean	SD	Mean	SD	1	2	3	4	5	6	7	8	9	10	11	12	13	14
1. Internal Integration	3.5	0.8	3.4	0.9		0.18	0.29	0.20	0.28	0.17	0.19	0.17	0.17	0.22	0.04	0.18	0.09	–0.16
2. External coordination	3.0	0.8	3.1	0.7	0.17		0.25	0.26	0.23	0.17	0.18	0.16	0.16	0.15	0.04	0.17	0.10	–0.17
3. Implementation	3.7	0.9	3.8	1.0	0.30	0.27		0.21	0.25	0.26	0.22	0.17	0.18	0.12	0.05	0.15	0.10	–0.13
4. Adaptation-in-use	3.6	0.9	3.7	0.8	0.21	0.25	0.22		0.12	0.25	0.23	0.20	0.18	0.19	0.03	0.15	0.11	–0.13
5. Change catalysis	3.3	1.2	3.4	1.4	0.29	0.25	0.27	0.13		0.29	0.25	0.22	0.23	0.18	0.04	0.10	0.16	–0.17
6. Operating performance	3.6	0.7	3.5	0.6	0.16	0.16	0.25	0.26	0.28		0.42	0.17	0.18	0.19	0.05	0.08	0.17	–0.18
7. ΔCost of Goods Sold [a]	0.04	0.13	0.05	0.18								0.17	0.16	0.15	0.08	0.05	0.07	–0.09
8. Δsales [a]	0.07	0.31	0.8	0.22									0.17	0.10	0.09	0.04	0.03	0.02
9. ΔGross Profit Margin [a]	0.05	0.22	0.06	0.16										0.04	0.10	0.03	0.03	0.04
10. Time since registration	40.5	17.9	43	15.6	0.21	0.15	0.11	0.20	0.18	0.19					0.17	0.21	–0.15	0.08
11. Number of employees [b]	184	7258	210	6258	0.05	0.03	0.05	0.03	0.04	0.05				0.18		0.05	0.05	0.05
12. Manufacturing / Service	0.25		0.23		0.17	0.18	0.16	0.15	0.11	0.08				0.20	0.05		0.04	0.03
13. External pressures	4.0	0.6	3.9	0.7	0.09	0.11	0.10	0.13	0.15	0.17				–0.16	0.05	0.04		0.05
14. Technological uncertainty	3.5	0.8	3.3	0.8	–0.15	–0.16	–0.12	–0.13	–0.15	0.17				0.07	0.07	0.03	0.06	

Correlations between facilities (full survey data) are below the diagonal, and correlations between companies (Compustat data) are above the diagonal.
$n = 1{,}150$ for facilities and 304 for organizations (Compustat data).
Correlations values equal to or greater than 0.17 are significant at 0.05 for the full survey data, and 0.20 for the Compustat data.
[a] Normalized.
[b] For the full sample, total number of facility employees, for the Compustat data total number of company employees ; Correlations are calculated for the natural logarithm of the total number of employees.

7.4 Results

The results of the OLS regressions appear in Table 7.2 below. We report standardized coefficients (betas; simple coefficients and standard errors are available upon request).

Hypotheses 1a and 1b predict that preparation will have a positive effect on implementation. The analyses corroborate these hypotheses, for both the full sample (model 2.1) and the public companies subsample (model 2.5). The coefficients of internal integration (H1a) and external coordination (H1b) are both positive and highly significant.

Hypothesis 2 predicts a positive interaction term between implementation and adaptation-in-use on performance. This hypothesis is strongly supported in predicting operating performance (models 2.4 and 2.8), as the interaction term is positive and highly significant. The significance level of this term is lower when predicting the three company-level performance measures (models 2.9–11). Yet, this interaction term is consistently positive and statistically significant. Therefore, the results support H2. Similarly, hypothesis 3 predicts a positive interaction term between implementation and change catalysis on performance. Again, for all four performance measures, the interaction term is positive and statistically significant, supporting H3. These results are especially encouraging due to the inherent difficulties in uncovering interaction effects in field studies (McClelland and Judd 1993).

It is interesting to examine the mediation effect of the two preparation measures by implementation, adaptation-in-use and change catalysis on performance. Essentially, the models in Table 7.2 allow us to construct a path-analysis model. However, due to the interaction effects (H2 and H3) the two preparation measures appear to have an intricate web of indirect effects. Both internal integration and external coordination have simple indirect effects mediated by their three succeeding variables (see Fig. 7.1). Yet, due to the interaction terms between these three succeeding variables the preparation measures have both indirect quadratic effects and interaction effects (internal integration by external coordination).[10] Because all relevant

[10]The formulation of the path analysis is as follows. (The following notation is used here: Internal Integration [II], External Coordination [EC], Implementation [IM], Adaptation-in-use [AD], Change Catalysis [CC], Performance [PR].)

$$\begin{aligned} PR &= a1 * II + a2 * EC + a3 * IM + a4 * AD + a5 * CC \\ &\quad + a6 * IM * AD + a7 * IM * CC \\ IM &= b1 * II + b2 * EC;\ AD = c1 * II + c2 * EC;\ CC = d1 * II + d2 * CE \end{aligned}$$

The first two element in the top equation are the direct effects of internal integration and external coordination on performance. Substituting the last three equations into the first one yields the following:

$$\begin{aligned} PR = &(a1 + a3 * b1 + a4 * c1 + a5 * d1) * II \\ &+ (a2 + a3 * b2 + a4 * c2 + a5 * d2) * EC + (a6 * b1 * c1 + a7 * b1 * d1) * II^2 \\ &+ (a6 * b2 * c2 + a7 * b2 * d2) * EC^2 + (a6 * b1 * c2 + a6 * b2 * c1 \\ &+ a7 * b1 * d2 + a7 * b2 * d1) * II * EC \end{aligned}$$

Specific numerical values for all models are available from the authors upon request.

Table 7.2 Results of regression analyses (standardized β coefficients)

	Model 2.1	Model 2.2	Model 2.3	Model 2.4	Model 2.5	Model 2.6	Model 2.7	Model 2.8	Model 2.9	Model 2.10	Model 2.11
Sample	Full Sample				Publicly-held sub-sample (Compustat data)						
Dependent variable	Implementation	Adaptation -in-use	Change catalysis	Operating perfor-mance	Implementation	Adaptation -in-use	Change catalysis	Operating perfor-mance	ΔCost of Goods Sold [a]	Δsales [a]	ΔGross Profit Margin [a]
Internal integration	0.47***	0.28**	0.26**	0.25*	0.37***	0.27**	0.26**	0.21*	0.09	0.08	0.09
External coordination	0.38***	0.24*	0.25*	0.23*	0.35***	0.25*	0.25*	0.29**	0.08	0.07	0.05
Implementation				0.29**				0.32***	0.06	0.09	0.11
Adaptation-in-use				0.28**				0.26**	0.04	0.05	0.04
Change catalysis				0.26**				0.24**	0.09	0.08	0.05
Implementation X adaptation-in-use				0.47***				0.39***	0.23*	0.26**	0.22*
Implementation X change catalysis				0.22*				0.28**	0.26**	0.24*	0.27**
Time since registration	0.21*	0.10	0.11	0.18*	0.25*	0.09	0.10	0.20*			

Table 7.2 (continued)

	Model 2.1	Model 2.2	Model 2.3	Model 2.4	Model 2.5	Model 2.6	Model 2.7	Model 2.8	Model 2.9	Model 2.10	Model 2.11
Sample	Full Sample				Publicly-held sub-sample (Compustat data)						
Dependent variable	Implementation	Adaptation -in-use	Change catalysis	Operating perfor-mance	Implementation	Adaptation -in-use	Change catalysis	Operating perfor-mance	ΔCost of Goods Sold [a]	Δsales [a]	ΔGross Profit Margin [a]
Number of employees	0.05	0.03	0.05	0.04	0.05	0.04	0.05	0.04	0.06	0.07	0.05
Manufacturing / service	0.04	0.04	0.05	0.03	0.04	0.04	0.04	0.03	0.05	0.04	0.06
External pressures	0.32***	0.08	0.09	0.24**	0.27***	0.07	0.11	0.26**	0.14†	0.15†	0.14†
Technological uncertainty	–0.23*	0.08	0.09	–0.17*	–0.19*	0.09	0.10	–0.16*	–0.13†	–0.12†	–0.11†
N	1139	1143	1141	1139	304	304	304	304	304	304	304
F	6.38***	4.23***	3.98***	5.97***	5.87***	3.98***	4.12***	6.38***	1.90**	2.52**	2.71**
R square	0.41	0.32	0.34	0.44	0.46	0.33	0.35	0.38	0.34	0.25	0.29
Adjusted *R* square	0.38	0.22	0.23	0.34	0.32	0.22	0.24	0.26	0.12	0.08	0.09

1-tailed t-test p-values
[a] – normalized.
† $p<0.1$
* $p<0.05$
** $p<0.01$
*** $p<0.001$

beta coefficients are positive, preparation turns out to have an amplified effect on performance via implementation, adaptation-in-use, and change catalysis.

We now turn to the effects of the control variables. Time since registration weakly increases the extent of implementation and of operating performance, but has no effect on either adaptation-in-use or change catalysis. Number of employees and industrial sector (manufacturing/service) seem to have no effect on either implementation or on any of the performance measures. External pressures for quality strongly increases extent of implementation and operating performance, but has only mild (positive) effect on financial performance. Last, firms operating in higher levels of technological uncertainty find it more difficult to implement the ISO 9000 standard. Such facilities and companies exhibit somewhat lower levels of implementation and lower performance levels. The effects of the control variables, in summary, are varied. The strongest effect is that of external pressures, which lead to deeper levels of implementation. None of the control variables affected either adaptation-in-use or change catalysis. Last, it appears that the obtained results are generalizable across establishment size and industrial sector.

7.5 Discussion, Limitations, and Implications

The study results highlight the importance of bridging two implementation gaps through learning in organizations. Preparation bridges the adoption–implementation gap by increasing the extent of implementation. This is a necessary but insufficient requirement for gaining from a change. "Action counts more than elegant plans and concepts" (Pfeffer and Sutton 1999: 251). However, preparation only ensures the *potential* of the change to produce performance gains. Learning-in-action bridges the implementation–performance gap. Adaptation-in-use, a form of first-order learning, perfects the implementation of a planned change. Change catalysis, a form of second-order learning, uses the current implementation activities as a springboard for additional innovation. In that, the combined effect of change catalysis and implementation suggests that implementation does not come to an end in perfecting an existing innovation, but is a part of an ongoing cycle of doing–reflecting–doing.

Our results also challenge the idea that exploitation and exploration are inherently at odds with each other. It is possible that process management innovations such as ISO 9000 do lead to more proximal technological innovation rather than path-breaking ones, as found by Benner and Tushman (2002). However, we take a more mundane perspective on the dual concepts of exploration and exploitation. As we examine these concepts in the context of reflective action (i.e., implementation), we see no innate contradiction between adaptation and change catalysis. Both make implementation more effective and both are driven by preparation activities.

Another perceived tension is between standardization and heterogeneity (Argote 1999). Conceivably, compliance with a common standard (in our case ISO 9000) should make organizations more homogenous. Yet, our study highlights important competitive differences between organizations. As Barney (1986) maintains, competitive advantage arises from the differences among firms (heterogeneity), but if all companies implement ISO 9000 in the same way, can a particular company derive a special benefit? Our study indicates that competitive advantage can be gained from implementing a common management practice if implementation is understood not as a discrete and homogenous industry-wide phenomenon, but when variations in this process are considered.

Our study is not free of limitations. Though we have taken measures to reduce common method variance (e.g., by triangulating performance data from our respondents and Compustat), we could not completely eliminate it. With limited resources, the choice was made to favor greater coverage over surveying multiple respondents per facility. We examined the implementation of a single administrative innovation – ISO 9000. Future research may uncover that different innovations may entail different implementation processes. Though we believe the concepts of learning-before-doing and learning-in-action are of importance in any context, an array of contingencies may change the balance between the two and the relative importance of adaptation-in-use and change catalysis. Such contingencies may be related to the type of innovation (administrative vs. technological), the novelty of the change (radical vs. incremental), and the cycle time of innovations in a particular setting. Moreover, this innovation has a built-in external auditing system. There might be a difference between audited and unaudited implementation efforts, when the audit may act as a self-reinforcing mechanism (Postrel and Rumelt 1992). Additionally, due to our selection criteria, our publicly held companies sub-sample was biased against large, multi-facility firms. In large, diversified firms, facility-level data become increasingly detached from firm-level performance. Conversely, our survey data came from a single source per facility, exposing our analyses to problems of causal inference and halo effects. Some comfort may be drawn from the substantial similarity in the analyses of the full (facility-level) sample and the public firms sub-sample.

Our findings impinge on three broad areas of interest for organization design theorists. First, the specific administrative innovation that was implemented, ISO 9000, bears a direct relation to a long-standing debate in organization theory about the effect of formalization and bureaucracy in general. Greater extent of ISO 9000 implementation indicates higher levels formalization. Hence, our findings support a rather unpopular view in organization design – higher-performing organizations are those with more, not less, formalization. Perhaps, to use Adler and Borys' (1996) terms, "enabling" bureaucracies (as opposed to the classic "coercive" ones) are more the norm than the exception. In such enabling bureaucracies, "formalization codifies best-practice routines so as to stabilize and diffuse new organizational capabilities" (Adler and Borys 1996: 69).

Second, at their core, these processes represent various performative and search routines in organizations. Our model provides an affirmation of the inherent capabilities to produce change, as argued by Feldman and Pentland (2003). Implementation is essentially a process of routinization. While the interrelations between implementation and adaptation-in-use are not surprising, the interrelations between implementation and change catalysis highlight the endogenous change aspect of routines – "change that comes from within organizational routines: change that is a result of engagement in the routine itself" (Feldman and Pentland 2003: 112).

Third, and more generally, our model dealt with processes and interactions between then. To understand organizations, one must strive to investigate the underlying processes (Sutton and Staw 1995). While substantial literature in organizational research presented these processes in their richness and complexity, our study took a different approach. These processes not only were quantified, but a model depicting these processes' interrelationship and effects on performance was theorized and tested. Hence, this study may serve as an example of depicting organizational processes in a less complex way, a way that allows modeling and quantifying these processes and their effects.

This study can provide direct implications for mangers. Both action and reflection are important for those who wish to manage change and to realize its potential benefits. Managers must not blindly insist on compliance to the normative model in their heads. Adaptation should not only be accepted, but encouraged. Similarly, the interplay between exploiting the proximal benefits of the current change and the more uncertain exploration must be recognized. In particular, the possibility that implementation opens the door for changes beyond the current one must not be ignored. Additionally, our focus on implementation and action does not make preparation activities redundant. Indeed, such learning-before-doing activities yield substantial dividends by making the implementation process more productive. This study is of particular use for managers who contemplate the benefits of ISO 9000. As mentioned earlier, there exists much debate about the benefits of this quality standard. We found that not all ISO 9000 registrations are the same. Even within ISO 9000-certified facilities there are considerable differences in the extent of implementation of the standard. Furthermore, ISO 9000 implementation can lead to substantial performance gains only when accompanied by adaptation, reflection, and readiness to extend the change beyond installation of this standard by itself.

7.6 Conclusion

In this research study, we found that indeed, luck does favor the prepared – thinking ahead about implementation facilitates implementation. Yet, this is not enough to leverage the implementation to effect performance. Learning-in-action must accompany implementation. Moreover, learning-before-doing also lays a ground fertile enough for learning-in-action, which makes implementation much more effective. Our findings also highlight the complex web of types of organizational processes – between those geared toward local search and adaptation and those related to exploring more distant opportunities. Indeed, among the most intriguing findings is that linking implementation of the current innovation with change catalysis – performance is enhanced by doing both. Apparently, exploration not only opens up new prospects, but also improves current performance.

Acknowledgments This chapter was supported in part from a grant from the NSF, Decision, Risk Analysis, and Management Division (SES-9905604). McGraw Hill Quality Systems and Plexus Corporation sponsored the survey of ISO 9000 registrants upon which this chapter is based. Gove Allen of the University of Minnesota designed the Internet site that was used to carry out the survey. Gove helped us gather additional data to supplement the survey from the Compustat data base. The authors wish to acknowledge Ayala Cohen, Shmuel Ellis, Martin Gannon, Jørn Flohr Nielsen, Roger Schroeder, and Charles Snow for their comments on this chapter and earlier drafts. This chapter also benefited from the comments of anonymous reviewers of the NSF proposal.

References

Abrahamson E (1991) Managerial fads and fashions: The diffusion and rejection of innovation. Academy of Management Review 16: 586–612.

Abrahamson E (1996) Management fashion. Academy of Management Review 21: 285.

Abrahamson E (1997) The emergence and prevalence of employee management rhetorics: The effects of long waves, labor unions, and turnover, 1875 to 1992. Academy of Management Journal 40: 491–533.

Adler PS, Borys B (1996) Two types of bureaucracy: Enabling and coercive. Administrative Science Quarterly 41: 61–89.

Aiken LS, West SG (1991) Multiple regression: Testing and interpreting interactions. Newbury Park, CA: Sage.

Aiman-Smith L, Green SG (2002) Implementing new manufacturing technology: The related effects of technology characteristics and user learning activities. Academy of Management Journal 45: 421–430.

Argote L (1999) Organizational learning: Creating, retaining, and transferring knowledge. Norwell, MA: Kluwer Academic Publishers.

Argyris C, Schön DA (1978) Organizational learning: A theory of action perspective. Reading, MA: Addison-Wesley.

Armstrong JS, Overton TS (1977) Estimating non response bias in mail surveys. Journal of Marketing Research 14: 396–402.

Bagchi TP (1996) ISO 9000: Concepts, methods, and implementation. 2nd edn. New Delhi: Wheeler Publishing.

Bardach E (1977) The implementation game: What happens after a bill becomes a law. Cambridge, MA: MIT Press.

Batchelor C (1992) Badges of quality. Financial Times, 1 September.

Barney J (1986) Organizational culture: can it be a source of sustained competitive advantage? Academy of Management Review 11: 656–665.

Barrett FJ (1998) Creativity and improvisation in jazz and organizations: Implications for organizational learning. Organization Science 9: 605–622.

Benner MJ, Tushman ML (2002) Process management and technological innovation: A longitudinal study of the photography and paint industries. Administrative Science Quarterly 47: 676–706.

Beyer JM, Trice HM (1978) Implementing change: Alcoholism policies in work organizations. New York: The Free Press.

Bossidy L, Charan R (2002) Execution: The discipline of getting things done. New York: Crown Business.

Brown JS, Duguid P (1991) Organizational learning and communities-of-practice: Toward a unified view of working, learning, and innovation. Organization Science 2: 40–57.

Brown JS, Duguid P (2001) Knowledge and organization: A social-practice perspective. Organization Science 12: 198–213.

Brown A, Loughton WT (1998) Smaller enterprises' experiences with ISO 9000. International Journal of Quality & Reliability Management 15: 273–285.

Brunsson N, Jacobsson B, Associates (2000) A world of standards. New York: Oxford University Press.

Carroll JS, Rudolph JW, Hatakenaka S (2003) Learning from organizational experience. In: Easterby-Smith M, Lyles MA (eds), The Blackwell handbook of organizational learning and knowledge management. Oxford, UK: Blackwell Publishing, pp 575–600.

Chelariu C, Johnston WJ, Young L (2001) Learning to improvise, improvising to learn. A process of responding to complex environments. Journal of Business Research 55: 141–147.

Cole RE (1999) Managing quality fads. New York: Oxford University Press.

Cooper R, Zmud R (1990) Information technology implementation. Management Science 36: 123–139.

Crossan MM, Lane HW, White RE (1999) An organizational learning framework: From intuition to institution. Academy of Management Review 24: 522–537.

Dierickx I., Cool K (1989) Asset stock accumulation and sustainability of competitive advantage. Management Science 35: 1504–1511.

Douglas TJ, Judge WQ Jr. (2001) Total quality management implementation and competitive advantage: The role of structural control and exploration. Academy of Management Journal 44: 158–169.

Edmondson AC, Bohmer RM, Pisano GP (2001) Disrupted routines: Team learning and new technology implementation in hospitals. Administrative Science Quarterly 46: 685–716.

Eisenhardt KM, Tabrizi BN (1995) Accelerating adaptive processes: Product innovation in the global computer industry. Administrative Science Quarterly 40: 84–111.

Elmuti D, Kathawala Y (1997) An investigation into the effect of ISO9000 on participants' attitudes and job performance. Production and Inventory Management Journal 38: 52–55.

Feldman MS, Pentland BT (2003) Reconceptualizing organizational routines as a source of flexibility and change. Administrative Science Quarterly 48: 94–118.

Fidler LA, Johnson JD (1984) Communication and innovation implementation. Academy of Management Review 9: 704–711.

Flynn BB, Schroeder RG, Sakikabara S (1994) A framework for quality management research and an associated measurement instrument. Journal of Operations Management 11: 339–366.

Gersick CJG, Hackman JR (1990) Habitual routines in task-performing groups. Organizational Behavior and Human Decision Processes 47(1): 65–97.

Greve HR, Taylor A (2000) Innovations as catalysts for organizational change: Shift in organizational cognition and search. Administrative Science Quarterly 45: 54–80.

Guler I, Guillen MF, Muir J (2002) Global competition, institutions, and the diffusion of organizational practices: The international spread of ISO 9000 quality certificates. Administrative Science Quarterly 47: 207–233.

Harman HH (1967) Modern factors analysis. Chicago, IL: University of Chicago Press.

Helfat CE, Raubitschek RS (2000) Product sequencing: Co-evolution of knowledge, capabilities and products. Strategic Management Journal 21: 961–979.

Hunt J (1997) Evaluation the tradeoffs: ISO9000 registration or compliance. Quality 36: 42–45.

ISO 9000 Survey (1996). New York: McGraw Hill.

James LR, Demaree RG, Wolf G (1984) Estimating within-group interrater reliability with and without response bias. Journal of Applied Psychology 69: 85–98.

Klein KJ, Conn AB, Sorra JS (2001) Implementing computerized technology: An organizational analysis. Journal of Applied Psychology 86: 811–824.

Klein KJ, Sorra JS (1996) The challenge of innovation implementation. Academy of Management Review 21: 100–1080.

Lave J, Wenger E (1991) Situated learning: Legitimate peripheral participation. New York: Cambridge University Press.

Lengnick-Hall CA (1996) Customer contribution to quality: A different view of the customer-oriented firm. Academy of Management Review 21: 791–824.

Leonard-Barton D (1995) Wellsprings of knowledge: Building and sustaining the sources of innovation. Boston, MA: Harvard Business School Press.

Lewis LK, Seibold DR (1993) Innovation modification during intra-organizational adoption. Academy of Management Review 18: 322–354.
March JG (1991) Exploration and exploitation in organizational learning. Organization Science 2: 71–87.
March JG, Schulz M, Zhou X (2000) The dynamics of rules: Change in written organizational codes. Stanford, CA: Stanford University Press.
Marcus AA (1988) Implementing externally induced innovations: A comparison of rule-bound and autonomous approaches. Academy of Management Journal 31: 235–258.
Matusik SF, Hill CWL (1998) The utilization of contingent work, knowledge creation, and competitive advantage. Academy of Management Review 23: 118–131.
McClelland GH, Judd CM (1993) Statistical difficulties of detecting interaction and moderator effects. Psychological Bulletin 114: 376–390.
Meyer AD, Goes JB (1988) Organizational assimilation of innovations: A multilevel contextual analysis. Academy of Management Journal 31: 897–923.
Miller CC, Cardinal LB (1994) Strategic planning and firm performance: A synthesis of more than two decades of research. Academy of Management Journal 37: 1649–1665.
Mirvis PH (1998) Variations on a theme: Practice improvisation. Organization Science 9: 586–592.
Moorman C, Miner AS (1998a) Organizational improvisation and organizational memory. Academy of Management Review 23: 698–723.
Moorman C, Miner AS (1998b) The convergence of planning and execution: Improvisation in new product development. Journal of Marketing 62: 1–20.
Naveh E, Marcus AA, Allen G, Moon HK (1999) ISO 9000 survey '99: An analytical tool to assess the costs, benefits and savings of ISO 9000 registration. New York: McGraw-Hill.
Nord WR, Tucker S (1987) Implementing routine and radical innovations. Lexington, MA: Lexington Books.
Nutt PC (1986) Tactics of implementation. Academy of Management Journal 29: 230–261.
Nutt PC (1989) Selecting tactics to implement strategic plans. Strategic Management Journal 10: 145–157.
Orlikowski WJ (2002) Knowing in practice: Enacting a collective capability in distributed organizing. Organization Science 13: 249–273.
Pfeffer J, Sutton RI (1999) The knowing-doing gap: How smart companies turn knowledge into action. Boston, MA: Harvard Business School Press.
Pisano GP (1996) Learning-before-doing in the development of new process technology. Research Policy 25: 1097–1119.
Postrel S, Rumelt RP (1992) Incentives, routines and self-control. Industrial and Corporate Change 1: 397–425.
Powell TC (1995) Total quality management as competitive advantage: A review and empirical study. Strategic Management Journal 16: 15–37.
Reger RK, Gustafson LT, Demarie SM, Mullane JV (1994) Reframing the organization: Why implementing total quality is easier said than done. Academy of Management Review 19: 565–584.
Repening NP (2002) A Simulation-based approach to understanding the dynamics of innovation implementation. Organization Science 13: 109–127.
Robin M, Dennis K (1994) An evaluation of the effects of quality improvement activities on business performance. The International Journal of Quality and Reliability Management 11: 29–45.
Rogers EM (1995) Diffusion of innovations, 4th edn. New York: The Free Press.
Schön DA (1983) The reflective practitioner. New York: Basic Books.
Sitkin SB, Sutcliffe KM, Schroeder RG (1994) Distinguishing control from learning in total quality management: A contingency perspective. Academy of Management Review 19: 537–564.
Staw BM, Epstein LD (2000) What bandwagons bring: Effects of popular management techniques on corporate performance, reputation, and CEO pay. Administrative Science Quarterly 45: 523–556.

Steiner GA (1979) Strategic planning. New York: The Free Press.
Sutton RI, Staw BM (1995) What theory is not. Administrative Science Quarterly 40: 371–384.
Szulanski G, Cappetta R (2003) Stickiness: Conceptualizing, measuring, and predicting difficulties in the transfer of knowledge within organizations. In: Easterby-Smith M, Lyles MA (eds), The Blackwell handbook of organizational learning and knowledge management. Oxford, UK: Blackwell, pp 513–534.
Terziovski M, Samson D, Dow D (1997) The Business value of quality management systems certification evidence form Australia and New Zealand. Journal of Operations Management 15: 1–18.
Thomas JB, Sussman SW, Henderson JC (2001) Understanding "strategic learning": Linking organizational learning, knowledge management, and sense making. Organization Science 12: 331–345.
Tushman ML, O′Rilley CA (1996) The ambidextrous organization: Managing evolutionary and revolutionary change. California Management Review 38: 8–30.
Wayhan VB, Kirche ET, Khumawala BM (2002) ISO 9000 certification: The financial performance implications. Total Quality Management 13: 217–231.
Weick KE (1998) Improvisation as a mindset for organizational analysis. Organization Science 9: 543–555.
Wenger E, McDermott R, Snyder WM (2002) Cultivating communities of practice: A guide to managing knowledge. Boston, MA: Harvard Business School Press.

Chapter 8
Underfits Versus Overfits in the Contingency Theory of Organizational Design: Asymmetric Effects of Misfits on Performance

Peter Klaas and Lex Donaldson

Abstract The contingency theory approach to organizational design traditionally treats underfit as producing equal performance loss as overfit, so that the effects of these misfits on performance are symmetrical. Recently, an asymmetric view has been proposed, in which underfit produces lower performance than overfit. This chapter analyzes these views. The effects of underfits and overfits on benefits and costs are distinguished. The differential effects on organizational performance of underfit and overfit are to be understood by their effects on benefit and costs. Implications are drawn for organization theory and design. For future empirical research, it is specified how to correctly identify differential performance effects of underfits and overfits. In a managerial design perspective, underfit is liable to occur in growing organizations and to rob them of some of their potential growth. While underfit will lead to an acute condition, overfit will be more chronic.

Keywords Contingency theory · Organization design · Fit · Misfit · Performance

8.1 Introduction

Within organizational theory, there has been a long-standing interest in the concept of fit and its relationship to organizational performance, and fit has served as an important building block for theory construction in research (Venkatraman 1989). The idea of fit was first introduced from biology in the 1950s. Fitness means that an arrangement seems to be useful for a certain purpose (v. Bertalanffy 1968: 77) and simply says that these are the kind of creatures that will "work," that is, survive, in this kind of environment (Simon 1982: 10). As an example of this thinking, an ice bear represents a fit with an arctic environment and hence it will survive, while a grizzly represents a misfit to that environment and would therefore not survive. Within the more narrow scientific scope of organization design (OD), many models share an underlying premise that context and structure must fit

P. Klaas (✉)
Vestas Assembly A/S, Rymarken 2, DK 8210, Aarhus V, Denmark
e-mail: klaas@mail.dk

A. Bøllingtoft et al. (eds.), *New Approaches to Organization Design*, Information and Organization Design Series 8, DOI 10.1007/978-1-4419-0627-4_8,

together if the organization is to perform well (Drazin and Van de Ven 1985). The basic concept of fit in organization design is the existence of one or several contingencies, with which one or more organizational parameters (e.g., organizational structure) must be congruent. Much research in OD has focussed on identifying fit relationships.

Early developments of fit research within OD includes "organic theory" (Donaldson 2001) or "contingency theory" (Scott 2003), which mainly focussed on the contingencies of task uncertainty, environment, and technology. This stream tended to use a prescriptive approach based upon the empirical results. Fit is treated as being an ideal (often) linear structure–context relationship; the farther away from the fit line (i.e., the farther away the structural parameter is from its ideal), the larger is the misfit and so the lower is the performance (Drazin and Van de Ven 1985).

Another early development was "bureaucracy theory" (Donaldson 2001) or "comparative structural analysis" (Scott 2003), focussing on organizational size as the contingency factor. The approach was primarily quantitative. Various structural parameters such as specialization, formalization, decentralization, and vertical span, vary systematically with organizational size. Misfit between these structural parameters and that required by organizational size is associated with lower performance (Child 1975).

Other developments within OD included the "gestalt" (Miller and Friesen 1980) or "configuration" (Doty et al. 1993) approach. Thus, a configuration is any multidimensional constellation of conceptually distinct characteristics that commonly occur together (Meyer et al. 1993: 1175). Fit is the internal consistency of multiple contingencies and multiple structural characteristics, assumed to affect performance characteristics (Drazin and Van de Ven 1985). Misfit is deviations from ideal-type designs; the larger the deviation, the larger is the performance penalty (Drazin and Van de Ven 1985; Venkatraman 1989).

One of the assumptions shared by the different misfit models has been that the performance effects from misfits are not affected by the direction of the misfit. However, when proposing a model for entrepreneurial firms' fit with their environments, Naman and Slevin (1993) acknowledged a difference between overfit and underfit, even if they did not conceptualize and test for this difference.

More recently, Klaas, Lauridsen and Håkonsson (2006) developed a theoretical model of asymmetric misfit, based on the information processing perspective on organization design. Tushman and Nadler (1978) have earlier suggested that underfit may be worse for organizational performance than overfit. Klaas et al. (2006) build on this insight to argue that underfit sacrifices the attainment of organizational goals, whereas overfit only incurs inefficiency while attaining organizational goals.

This argument that underfit is worse for performance runs contrary to a long tradition in contingency theory, which treats underfit as producing the same performance as overfit. Its implications for organizational design and managerial practice are that greater efforts should be directed toward eliminating, or reducing, underfit than overfit. Moreover, theoretical reasons lead to the expectation that underfit will tend to occur in organizations that are growing. Underfit therein works to reduce the growth rate. Hence managers of growing organizations would be well advised to eliminate or avoid underfit. These are substantial changes to the theory and practice of organizational design, and therefore warrant a careful dissection of the arguments.

Thus, this chapter seeks to offer an elucidation of the idea of asymmetric effect of misfit on performance. The chapter begins with an explication of the traditional view that the performance effects of misfit are symmetric for underfit and overfit. We then consider the counter-view, of asymmetry, put forward by Klaas et al. (2006). We seek to explicate this view in the remainder of the chapter. In the final part of the chapter, we discuss the implications of the asymmetry view for empirical research and for the practice of organizational design.

8.2 Symmetric Effect of Misfits on Performance

In organizational design, there is a strong interest in the effect of organizational structure on organizational performance. The contingency theory of structure holds that the fit of organizational structure to the organizational contingency produces higher performance (Donaldson 1996, 2001). For example, the fit of specialization to the organizational size contingency produces higher performance (Child 1975). An organization with a different level of the structural variable would be in misfit, which causes lower performance. The greater the misfit, the lower is the performance, as a result. Misfit can occur in two ways. The structural level can be too high to fit the contingency, which is termed "overfit." Or the structural level can be too low to fit the contingency, which is termed "underfit." Traditionally, both kinds of misfit are considered to have the same effect, lowering performance to the same extent. The greater is the degree of misfit, either of overfit or underfit, the lower is the performance. *A one degree of overfit produces the same amount of performance loss as a one degree of underfit*. This is the theory that the effects of overfit and underfit are symmetrical. Klaas et al. (2006: 159) refer to this as the "balanced misfits" view – because each type of misfit is equal in its effect on performance.

Fit may be conceived of as a line, in which, for each level of the contingency variable, there is a corresponding level of the structural variable that fits it (Donaldson 2001). Usually, the fit line is considered to be a diagonal line in the space formed by the contingency variable as the horizontal axis and the structural variable as the vertical axis (Drazin and Van de Ven 1985). For instance, for size, as the contingency on the horizontal axis, and formalization, as the structural variable on the vertical axis, the fit line runs diagonally from bottom left to top right, that is, from low size and low formalization to high size and high formalization. The greater is the contingency variable, the greater is the level of the structural variable that is required to fit it. For instance, the larger is the size of an organization, the higher is the level of formalization needed to fit it. In Fig. 8.1, contingency and structure are both measured by five point scales: from 1 to 5. Fit is where the structural (e.g., formalization) level matches the contingency (e.g., size) level. Organizations that are not on the fit line are in misfit. Those above the fit line are in overfit. Those below the fit line are in underfit.

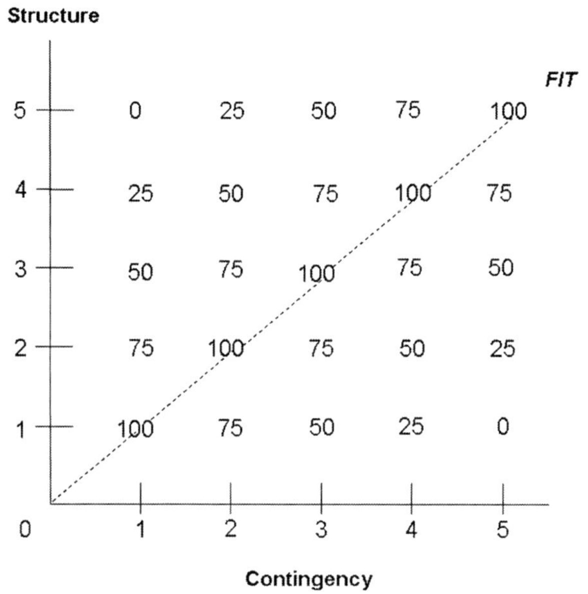

Fig. 8.1 Symmetric misfit and performance

If, to the two dimensions of contingency and structure, performance is added, as a third dimension, then the fit line is like the ridge that runs along the top of the roof of a house. The misfits are like the eaves of the house, which slope downwards. Because the effect of misfit on performance is the same for overfit and underfit, the gradients of the slopes are the same for overfit and underfit (Fig. 8.2).

The example shown in Figs. 8.1 and 8.2 is hypothetical, in which performance is measured in "performance units," that is, the degree to which organizational goals are attained (e.g., dollars for a business firm or number of patients receiving quality care in a hospital). Organizations that are in fit attain the highest level of performance, 100 performance units (p.u.), and lie on the fit line. Each step away from the fit line increases misfit by one degree and reduces performance by 25 p.u.

The effect of misfit on performance is considered to be symmetrical because the outcome of misfit is the same whether it is above or below the fit line. More precisely a one degree of misfit (i.e., one structural level away from the fitting structural level) causes the same performance loss whether it is an overfit or an underfit. For instance, in Fig. 8.1 the fit to a contingency level of 3 is a structural level of 3. A misfitting structural level of 4, that is, overfit of one structural level, causes the same performance loss as a misfitting structural level of 2, that is, underfit of one structural level. Both structural levels 4 and 2, being misfits of one degree, have performances of 75 p.u., that is, 25 p.u. less than fit, which has performance of 100 p.u.

Failure to fit the contingency level, whether overfit or underfit, causes the same amount of loss of organizational performance. Woodward's (1965) pioneering contingency theory study showed that organizations in misfit, whether above or below the structural level that fitted the technology contingency, tended to be of

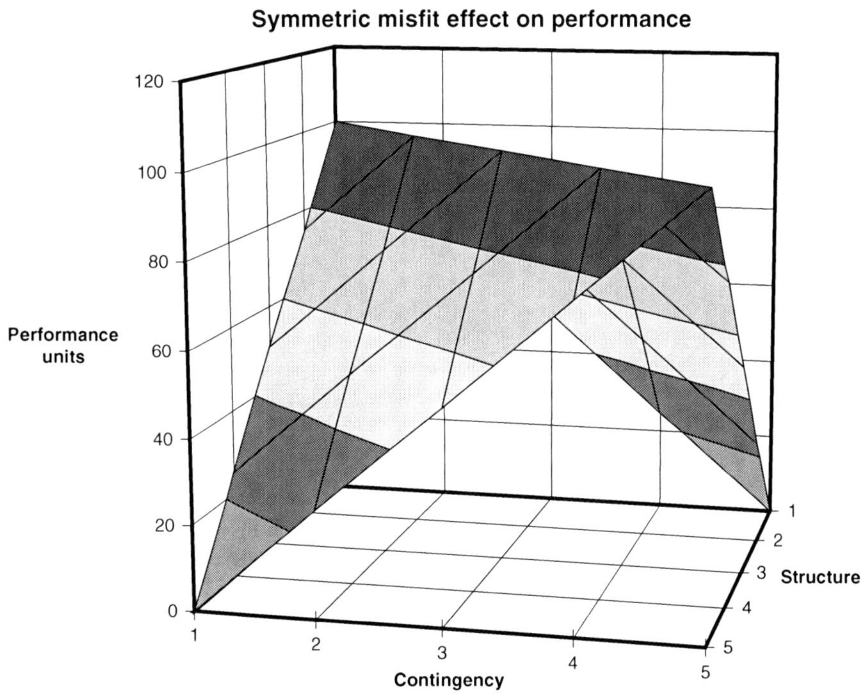

Fig. 8.2 Symmetric misfit effects on performance

lower performance, which is consistent with the idea of symmetrical effects of overfit and underfit. Similarly, the deviation approach of Keller (1994) predicts that a one degree of overfit causes the same performance loss as a one degree of underfit. Again, the profile deviation score approach using the Euclidean distance measure of misfit (Van de Ven and Drazin 1985) treats overfit and underfit as causing the same performance loss. Thus, the idea of symmetry of misfit effects of overfit and underfit is quite well established in the empirical literature on organizational design research and associated methodology, including those examining multiple misfits (Burton, Lauridsen and Obel 2002) and misfit of culture (Burton, Lauridsen and Obel 2004).

The underlying idea is that failure to fit, by either overfit or underfit, leads to performance loss of the same amount. Yet there is not necessarily any well-developed theory to undergird this view. It does not seem to be inherent in contingency theory that the effects of overfit and underfit are symmetric.

8.3 Asymmetric Effect of Misfits on Performance

The effect of misfit may be not symmetric, but asymmetric, so that there are two different effects of misfit: of overfit and underfit. Previously, Zajac, Kratz and Bresser (2000) have distinguished between underfit and overfit, though for strategies and

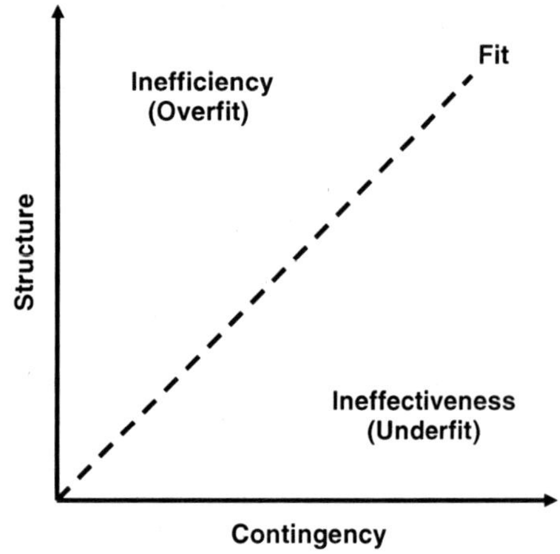

Fig. 8.3 The different concepts of over and underfit

their fit to the situation, rather than the fit of organizational structural designs to their contingencies, which is the concern here.

Klaas et al. (2006) have recently challenged the traditional assumption of symmetric effects of overfit and underfit. Instead, they argue that the effect of misfit on performance is asymmetric. Specifically, Klaas et al. (2006: 159) argue that, while both overfit and underfit cause performance loss, underfit causes more performance loss than overfit: "...while any misfit causes performance decreases, underfit is worse for performance than overfit."

Their argument is that attaining fit leads to effectiveness, in the sense that the organization will attain its goals (such as profitability). Thus, as underfit is reduced by the organization increasing its structural level, it moves up closer to, or onto, the fit line from underneath it, thereby increasing effectiveness (see Fig. 8.3). Once the organization is on the fit line, any further increase in the structural level, so that the organization rises above it into overfit, produces no further increase in effectiveness.

There is, however, typically also a cost to the organization in increasing its structural level, so that going above the fit line increases costs without increasing effectiveness. For overfit, the effectiveness produced is the same as being on the fit line, whereas the cost is greater than on the fit line. Hence, Klaas et al. (2006) argue overfit leads to inefficiency, in that the same effectiveness is attained but at greater cost. The organization in overfit is effective but inefficient. The overall performance is therefore less than when in fit.

Overall, decreasing underfit causes greater effectiveness, whereas decreasing overfit causes greater efficiency. Where effectiveness is defined as the degree to which the organization attains its goals, while efficiency is defined as the ratio of output to input, which is to say, the cost of attaining any degree of goal attainment. Thus, increasing underfit causes greater ineffectiveness, whereas increasing overfit causes greater inefficiency. Because ineffectiveness causes more performance loss than inefficiency, underfit causes more performance loss than overfit.

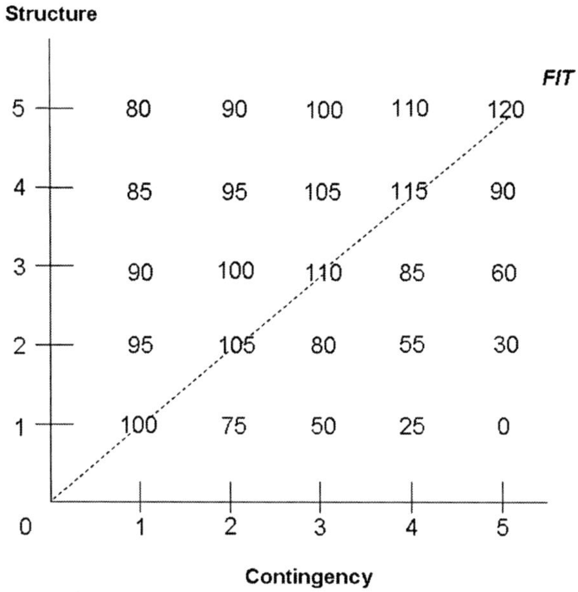

Fig. 8.4 Asymmetric misfit and performance with hetero-performance

To put it another way, since effectiveness contributes more to organizational performance than efficiency, then the performance of an organization in overfit is more than the performance of an organization in underfit. Hence the effect of misfit on performance is asymmetrical.

Figures 8.4 and 8.5 illustrate this asymmetric relationship between misfit and performance, where underfit has lower performance than overfit. Each degree of underfit produces 30 p.u. less of performance, whereas each degree of overfit produces only 5 p.u. less of performance. The "eaves of the roof of the house" slope more steeply downwards for underfit than for overfit (Fig. 8.5).

Effectiveness is modeled herein in terms of benefit. Benefit is the degree of attainment by an organization of its goals. As an example, a firm's goal may be profit maximization. An optimal structure assists profit maximization, such as by ensuring responsiveness to changing customer taste, through coordinating sale and product design functions, and providing economical production. Thereby the structure produces benefit to its organization.

The optimal structure will also be no more costly than necessary. Previously, Child (1972) has identified cost of structure, as distinct from its performance, as a consideration for managers in choosing between structures. The costs of a structure are those of designing, implementing, adjusting, and maintaining it. For example, for a decentralized structure, costs include recruiting and training managers able to handle the decisions that are being delegated to them from the CEO, and the costs of distributing information to them through the IT system. For a formalized structure,

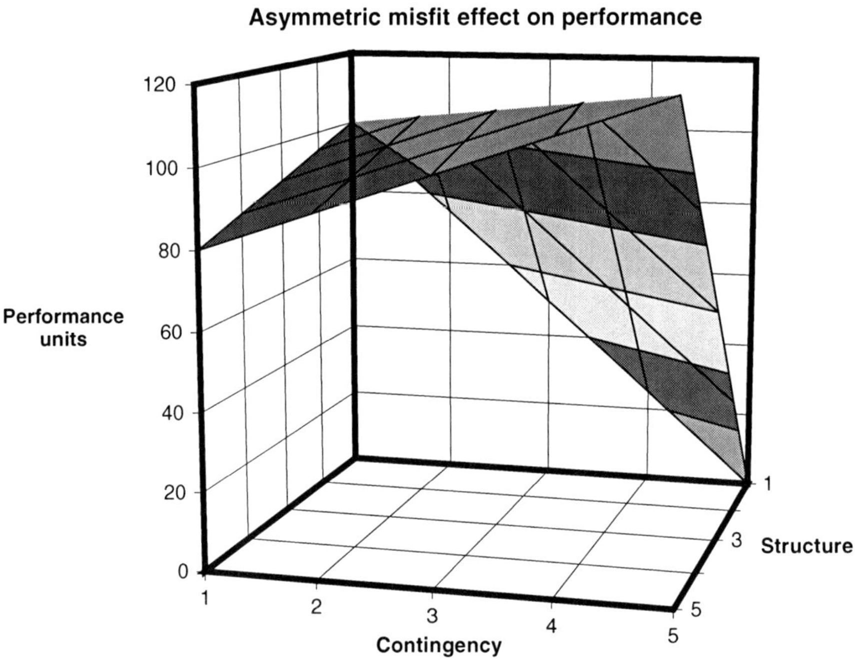

Fig. 8.5 Asymmetric misfit effect on performance

meaning a structure that uses rules and standard operating procedures (SOPs) to regulate the behavior of its members, costs include formulating and disseminating such rules and SOPs, training employees in them, and monitoring employee compliance. As a structural variable increases so too does its cost.

One element of the costs of structure is the increase in administrative personnel that is associated with higher levels of the structural variable. These administrative personnel incur a cost that is part of the overhead costs of an organization. As firms increase the level of their divisionalization, by going from functional to divisional structures, so the ratio of administrative to production personnel increases. In Grinyer and Yasai-Ardekani (1981: 478, Table 2) divisionalization is positively correlated with the ratio of administrative to production personnel, +0.3. Therefore, a one standard deviation increase in divisionalization increases this administrative ratio by 0.3 of a standard deviation. Clearly, such an additional amount of administrative personnel relative to production personnel imposes a cost on an organization and increases its overheads. This increase in administrative costs flows from the greater structural complexity of divisional relative to functional structures and shows that higher levels of a structural variable can be more costly. Inefficiency is modeled herein in terms of structural costs.

In underfit, an organization is ineffective, in that its benefit is less than the maximum, while its inefficiency is low, in that the cost of its structure is low. Because benefit affects performance more than structural cost does, performance in underfit is low. In overfit, an organization is effective, in that its benefit is at the maximum, while its inefficiency is high, in that the cost of its structure is high. But because benefit affects performance more than does structural cost, performance in overfit is higher than performance in underfit.

The asymmetric effect of misfit on performance can be likened to a person buying a jacket. Whether a jacket is good or bad depends upon whether or not it fits the person. A good jacket fits the person's size. If the jacket does not fit the person's size then it is not an optimal jacket. If the jacket underfits, it is too small for the person, who may not even be able to get inside the jacket, so it is a useless jacket. Instead, if the jacket overfits, it is too big for the person, so it looks unsightly, but the person is still able to get inside the jacket, and wear it, so it is still a useful jacket; it just has more capacity than that person needs. Hence, underfit is worse than overfit. This would hold even if the larger jacket cost more than smaller jacket, because it needs more material.

8.4 Information-Processing Theory

Klaas et al. (2006) base their theory on the information-processing view of organizations (Galbraith 1974; 1977; Tushman and Nadler 1978). In the information-processing view of organizational structure, the optimal design is that which processes the information that is necessary for the organization to attain its goals. Thus, the organization is in a situation which makes a demand for information-processing and ideally the organization also possesses a structure that has the capacity for processing that amount of information. In the optimal design, the IP demand is met by the IP capacity. Sub-optimality results from either the IP capacity being insufficient for the IP demand or from the IP capacity exceeding the IP demand.

Relating the information-processing view to structural contingency theory, the contingencies define the IP demand and the structure provides the IP capacity. In Klaas et al. (2006), the contingency is environmental complexity, uncertainty, and equivocality. In the optimal design, the IP demand from the contingency(ies) is met by the IP capacity supplied by the structure. Sub-optimality results in two distinct ways. If the IP capacity from the structure is insufficient to meet the IP demand from the contingencies, there is underfit. If the IP capacity from the structure exceeds the IP demand from the contingencies, there is overfit.

As Tushman and Nadler (1978: 619) state:

> ...the basic design problem is to balance the costs of information-processing capacity against the needs of the subunit's work – too much capacity will be redundant and costly; too little capacity will not get the job done.

In terms of the information-processing view, organizational structure produces the benefit from providing IP capacity, but organizational structure incurs a cost. When the structure fits the contingencies, the IP capacity of the structure matches the IP demand from the contingencies, producing the benefit of goal attainment with no more than the necessary cost. However, in misfit, the benefit and cost of the structure are out of kilter. In underfit, the IP capacity of the structure is insufficient to match the IP demand from the contingencies, so that the goals are not attained. The result is a lack of benefit, despite low cost, producing too low a benefit relative to cost. The organization is ineffective, in that it is failing to attain its goals, because of failure to invest in structure. Conversely, in overfit, the IP capacity of the structure more than matches the IP demand from the contingencies, so that while the goals are attained, the cost is excessive, producing too low a benefit relative to cost. The organization is inefficient, in that it is attaining its goals, but at too high a cost, because of over-investing in structure.

8.5 Benefits and Costs in Underfits and Overfits

In order to analyze more deeply asymmetric misfits and their relationship with performance, it is necessary to identify the separate costs and benefits, and how they vary between, and within, underfits and overfits. Thus, we need to take a slice through Fig. 8.4 and show the underlying costs and benefits that together determine performance. (We will do this for contingency level 3, because, being the middling position, this will most typify the general situation.)

Performance is benefit from the structure minus cost of the structure. For an organization below the fit line, that is, in underfit, the farther it is from the fit line, the less is its performance. The closer that an organization in underfit moves up toward the line, the more its performance increases. This increase in performance is the net of two effects. As the organization increases its performance, it receives more benefit from that increased level of the structural variable. However, as the organization increases its performance, it also incurs more cost from that increased level of the structural variable. Given that the benefit from increasing structure is more than the cost of increasing structure, then the net effect (benefit minus cost) of increasing structure is greater performance.

For an organization above the fit line, that is, in overfit, the farther it is from the fit line, the less is its performance. Again, its performance is the net of two effects: benefits and costs. For an organization in overfit, the effect of structure on benefit is constant, that is, benefit plateaus. The organization attains maximum benefit when in fit, and as the structural variable increases into overfit so the resulting benefit is at that same, maximum level. This is because effectiveness does not decrease as the organization moves beyond fit into overfit, because its structure is able to process all the information necessary to attain the organization's goals. However, there is increasing inefficiency as overfit increases, meaning that costs increase as the structural level increases in overfit. Hence, as the organization moves from fit into increasing overfit, benefit remains constant while cost increases. Therefore, increasing overfit means that benefit minus cost decreases with increasing structure, so that performance declines with increasing structure.

We may put it more generally in a formal model (Fig. 8.6) to show that underfit produces less performance than overfit.

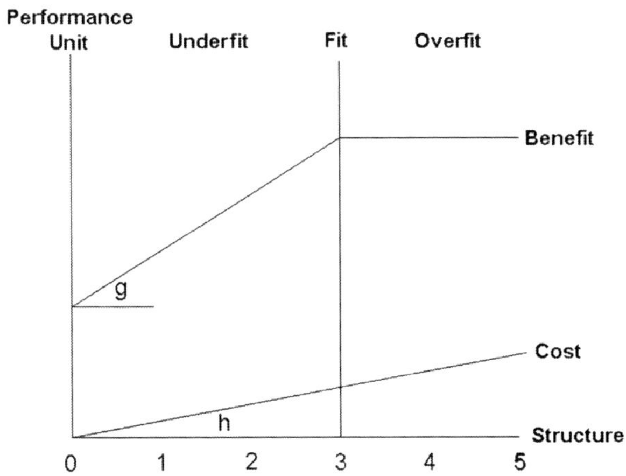

Fig. 8.6 Slopes of benefit and cost in asymmetric theory

The condition for which the performance of an underfit is inferior to that of the overfit (in the same degree of misfit) is as follows. The slope of the change in structural cost as the structural levels increase with the contingency level is h, and this slope is constant from underfits through overfits. One degree of increase in the structural level, and hence change in misfit, produces h amount of increase in structural cost. The slope of the change in benefit as the contingency level increases from underfit to fit is g, where after the benefit plateaus at its maximum value and so becomes constant. Both h and g are measured in the same units. Comparing one degree of underfit with one degree of overfit, underfit has g less benefit but saves $2h$ in cost. Therefore, if g is more than $2h$, then underfit has worse performance than overfit. Thus, underfit is worse than overfit if g is more than twice h. For two degrees of misfit, underfit is worse than overfit if $2g$ is more than $4h$, which equals g is more than $2h$. Hence the relationship generalizes for any degree of misfit: underfit is worse than overfit if g is more than twice h. Underfit is always worse than overfit, if the increase in benefit from decreasing underfit (g) is more than twice the increase in cost of each structural level (h). Wherever this condition holds, an underfit performs lower than the overfit of the same degree of misfit. Hence asymmetric effect of misfit on performance applies as long as structural cost increases at less than half the rate that benefit increases as underfit reduces.

For any level of the contingency variable, performance is highest at fit. However, fits to lower levels of the contingency variable occur at higher levels of the structural variable. Because costs rise with structural level, if benefits were constant across contingency levels, fits to higher levels of the contingency variable would have lower performances than fits to lower levels of the contingency variable. This implies that an organization would sacrifice performance if it moved to a fit at a higher level of the contingency. Therefore, rational organizations and their managers would never move beyond the lowest level of the contingency variable, which is contradicted by the fact that organizations exist at these higher levels (e.g., larger size), because they have increased their level of the contingency variable. Hence the model being proposed herein has to make such growth feasible – at least for some contingencies. Hetero-performance (Donaldson 2001) holds that fits to higher levels of the contingency variable have higher performance than fits to lower levels of the contingency variable. Thereby, organizations have an incentive to increase their level of the contingency variable. Given a positive main effect of contingency on performance that is greater than the decline in fit performance with contingency caused by the asymmetric effects of misfit, performance would *rise* with the contingency variable, that is, hetero-performance would be attained. This main effect of contingency on performance is conceptualized here as a positive effect of contingency on benefit which entails no costs and so flows through to raise performance.

The rate at which the performance of fits decreases with increases in the contingency level is h (because the decrease is caused by the increasing cost of the fitting structural level.) Thus, for hetero-performance (i.e., rising performance of fits to higher levels of the contingency), the contingency variable needs to raise the performance of fits by more than h for each increase in the contingency level. For size, the number of members of an organization (e.g., employees of a business firm), has an effect as each member is able to process information, and so size contributes positively to the information processed by the organization. Similarly, technology (e.g., computers and e-mail) contributes positively to information-processing by the organization. Hence, contingency variables such as size and technology can be conceived of as contributing positively to performance.

This postulate of a contingency effect on performance to provide an incentive, is only required, however, where organizational management have a choice of whether to increase contingency levels or not, which applies to internal organizational contingency variables. These contingencies include size, strategy, and technology. Other, extra-organizational contingencies, such as the environment, are not under the control of the organization's management, therefore an organization could be at a higher level of such a contingency, for example, high environmental uncertainty, without any incentive. Hence, theoretically, we expect that the asymmetric effect of misfits on performance (underfit worse than overfit) is accompanied by a positive main effect of the contingency on performance for intra-organizational contingencies (e.g., organizational size), but not for extra-organizational contingencies (e.g., the environment).

As an example, consider formalization and its fit to size. We will first show how hetero-performance can be created and then show how underfit can lead to lower performance than overfit. Formalization and size are both measured on five point scales. The fit line is where the level of the formalization variable equals the level of size. The fit line (see Fig. 8.4) exhibits hetero-performance, in that performance increases along the fit line from 100 to 120 p.u., so that fits to successively higher contingency levels are 5 p.u. higher than the fit point immediately below. This is because the benefit of each successive fit point is 10 p.u. higher than the fit point immediately below, while the cost is 5 p.u. higher than the fit point immediately below (because the fit is a higher level of the structural variable and costs rise by

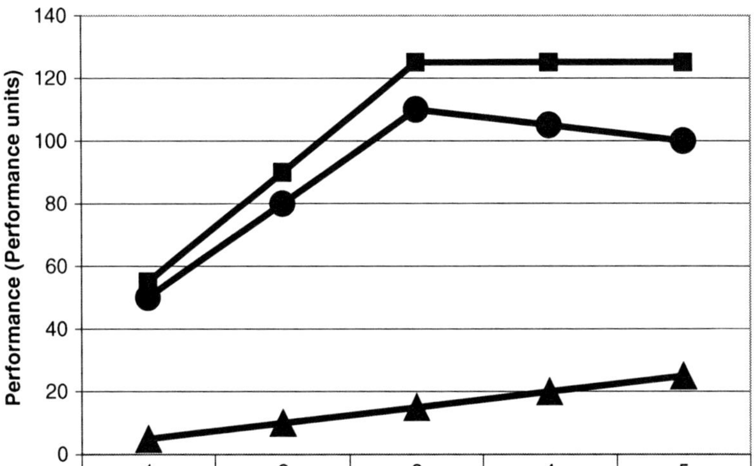

	1	2	3	4	5
Benefits	55	90	125	125	125
Costs	5	10	15	20	25
Performance	50	80	110	105	100

Structural level

Fig. 8.7 Performances of underfits and overfits from benefits and costs

5 p.u. for each structural level). The cost of fit is not from the fit itself but because of the structural cost of the fit.

We will now show how underfit leads to lower performance than overfit.

For size of level 3 (Fig. 8.7), the fitting level of formalization is 3, so that formalization of level 2 is one degree of underfit and formalization of level 4 is one degree of overfit, that is, the same degree of misfit. An underfit of formalization level 2 produces a benefit of 90 p.u. at a cost of 10 p.u. (because cost increases by 5 p.u. for each formalization level). Hence the performance of underfit of formalization level 2 is 80 p.u. (see Fig. 8.7). An overfit of formalization level 4 produces a benefit of 125 p.u. at a cost of 20 p.u. Therefore, the performance of overfit of formalization level 4 is 105 p.u. The result is that underfit (80 p.u.) has lower performance than overfit (105 p.u.). Similarly for misfits of two degrees: the performance of underfit is 50 p.u. (= benefit of 55 p.u. – cost of 5 p.u.), which is lower than the performance of overfit, 100 p.u. (= 125 p.u. – 25 p.u.) (Fig. 8.7). Underfit produces less performance than overfit of the same degree of misfit. The mean performance of underfits is 65 p.u., which is less than the mean performance of the overfits, 102.5 p.u.

To put it another way, if performance *loss* is the performance decrement from *not* being in fit, the performance loss from a degree of underfit is greater than the performance loss from a degree of overfit. Though cost is greater for overfit than underfit, overfit has the advantage of benefit being at its maximum value, whereas underfit has less than the maximum value of benefit. This illustrates the point of Klaas et al. (2006), that the ineffectiveness from underfit is more damaging than the inefficiency from overfit.

Thus, we see an illustration of how asymmetric effects of misfit on performance can arise from asymmetries in costs and benefits. The cost of the structure rises from underfit to overfit. The benefit also rises from underfit to overfit, and more steeply than the cost rises, so that performance increases from underfit to overfit. This is a formal modeling of the asymmetry theory of Klaas et al. (2006).

8.6 Differences in Degree of Misfit Between Underfits and Overfits

Our focus in this article is on the effects on performance being different between underfits and overfits. These differences should be carefully distinguished from other types of asymmetry between underfits and overfits. In a symmetrical distribution about the fit line, organizations would be equally distributed so that there were equal numbers of organizations above and below the fit line. Moreover, in a symmetrical distribution, the degrees to which the organizations were in misfit would be the same between underfits and overfits. Thus, the average degree of misfit would be the same between underfits and overfits.

Even if the distribution of organizations was symmetrical, so that there were equal numbers under and above the fit line, the mean misfits could still differ between underfits and overfits, because the organizations were scattered farther away on one side of the line than the other side. The side with the greater degree of average misfit would thereby also have the lower performance (other things being equal). Even if the effect on performance were the same between underfit and overfit, a greater degree of misfit among underfits would produce lower performance of underfits than overfits. Thus, underfits can suffer lower performance than overfits just by being more in misfit. If, over time, organizations in underfit strayed farther into misfit, or stayed there longer and adapted into fit less than organizations in overfit, then underfits would perform less than overfits.

In a previous section we stressed, from a methodological point of view, the desirability of separating out differences in performance that are due to the different effects on performance of underfits and overfits, from those due to differences in degrees of misfit. However, they could be connected theoretically. The structural adaptation to regain fit model (SARFIT) (Donaldson 1987, 2001) holds that organizations in misfit tend to remain in misfit until a crisis of low performance propels them to adapt into fit. Given that underfit produces lower performance than overfit, organizations in underfit are more likely to suffer a performance crisis and adapt into fit, thus reducing the average degree of misfit of underfits compared to overfits, as well as the number of organizations in underfit. In contrast, an organization in overfit will have less performance loss from overfit, so that they stray farther into misfit and will remain in misfit longer before adapting to fit; hence their average

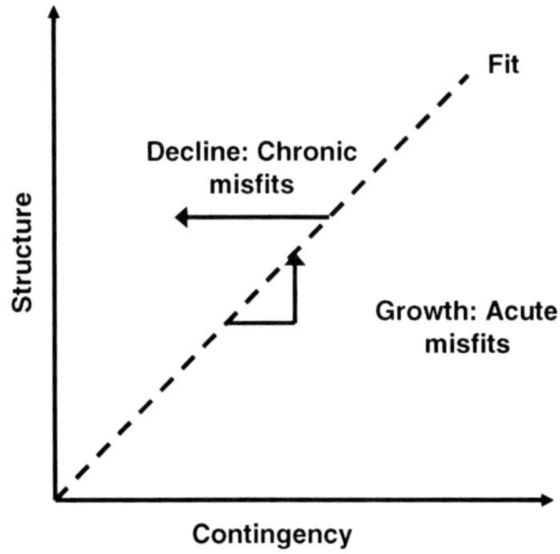

Fig. 8.8 Misfit behavior in growing and declining organizations

misfit will be more than for underfit and there will be more organizations in overfit than underfit.

It may be that we should distinguish between acute and chronic misfits (see Fig. 8.8). An acute misfit produces performance so poor that it triggers a crisis that leads to adaptive change to remedy the problem. Underfit would tend to be acute, for the reasons stated. Chronic misfit produces low performance, without being so poor that it triggers a crisis. Hence there is no adaptive change and the organization remains in misfit for a considerable period, suffering degraded performance. Overfit would tend to be chronic, for the reasons stated.

Thus, there can be theoretical linkage between the three asymmetries being discussed in this section: the effect of misfit types (underfit versus overfit) on performance, the average misfit and the number of organizations in the misfit types. While underfit causes more performance loss than overfit, overfits are on average in worse misfit, which is another route to depressing their performance. Because there may be more organizations in overfit than underfit, the cumulative effects of mean misfits and numbers in misfit types could be that the injurious effect on the performance of organizations across the economy would be more equal between underfit and overfit, despite underfit causing lower organizational performance than overfit.

8.7 Implications for Organization Theory and Practice

The theory of asymmetric effects of misfit holds that the effects on performance of overfit and underfit differ. An empirical result finding that they have the same effects would be support for the effects being symmetric, and hence support the traditional view. However, if, instead, the effect of underfit on performance is stronger than the effect of overfit on performance, then this would support the asymmetry theory. In this way, the empirical validity of the divergent views can be ascertained and the most valid view identified for a data set. The relationships could be tested by comparing the underfits and overfits by their means, slope coefficients, or correlation coefficients.

8.7.1 Means

Perhaps the simplest test of the theory that underfit performs lower than overfit is to compare the mean performances of the organizations in underfit with those in overfit. The hypothesis would be that the mean performance of the organizations in underfit would be lower than the mean performance of those in overfit.

However, this would be a valid comparison only if the degree to which the underfits were in underfit was the same as the degree to which the overfits were in overfit, that is, both groups were in misfit to an equal degree. Otherwise, if one group were more in misfit than the other, that in itself would lead it to have a lower mean performance.

The distribution of organizations either side of the fit line may not be symmetrical. There may be more organizations in misfit on one side that the other, for example, more in underfit than overfit. But, more importantly for a comparison of means, the degree of misfit may be different, for example, the average underfitted organization may be more in misfit than the average overfit. This latter will affect the comparison of the mean performances of underfit versus overfit. This comparison will reflect both the strength of the different misfit-performance relationship between underfits and overfits and their degrees of misfit. Thus, even if the effect of misfit on performance is asymmetrical, that is, overfit and underfit have different inherent effects, overfit and underfit could appear to have the same effect if their average misfits differ. For instance, if the average overfit is greater than the average underfit, then the effect of underfits and overfits could appear to be the same, even though underfit causes more performance loss than overfit.

As just stated, any difference between underfits and overfits in their degrees of misfit can be captured by the average (mean) misfit of the underfits and the average misfit of the overfits. A correction factor could be applied by taking the ratio of the means of the misfits of the two groups. This ratio can then be applied to the mean performances of one of the groups, before then comparing that with the mean performance of the other group. For instance, if overfits had a mean misfit that was twice as high as the underfits, then the mean performance of overfits should be corrected by halving it, before then comparing it with the mean performance of the underfits.

8.7.2 Slope Coefficients

The asymmetric theory holds that misfit causes more performance loss for underfits than overfit. Therefore the slope of performance on misfit should be stronger for underfits than overfits. Using the roof analogy, the slope of the eaves is greater on the underfit than the overfit side of the roof ridge (i.e., the fit line). Therefore, the slope coefficients of performance on misfit could be compared for underfits and overfits – the prediction being the slope would be more negative for underfits than overfits.

The slope coefficient can be subject to attenuation, however, if the variation of misfit or performance is restricted, so that the slope is reduced in value. As discussed below, for correlations, bias from attenuation can be avoided by applying a simple correction. Because the standardized slope coefficient is identical to the correlation coefficient, the procedure discussed below for correlations can be applied to standardized slope coefficients. For unstandardized slope coefficients, they need to be multiplied by the ratio of the standard deviations of misfit and performance, to make them equal to correlations, before applying the procedure to correct correlations for differential attenuation.

8.7.3 Correlation Coefficients

A comparison may be made of the coefficients of correlation between misfit and performance for the underfits and overfits. The hypotheses would be that the negative correlation between misfit and performance would be stronger for underfits than overfits.

The correlation of misfit with performance is attenuated (i.e., reduced) if the variation in the degrees of misfit is smaller, leading to under-estimates of the true effect of misfit on performance. Therefore, if there were differential attenuation between underfits and overfits, it would disturb the observed effects of overfit and underfit, possibly leading to a false conclusion, for example, that they are the same when they actually differ. Hence, if underfit was more attenuated than overfit, this would reduce the correlation coefficient of underfit more than any reduction for overfit, so that the correlation between underfit and performance could, falsely, appear weaker than the correlation between underfit and performance, despite the underfit correlation really being stronger than the overfit correlation.

The degree of attenuation of correlation is a function of the standard deviation of misfit, for both overfit and underfit. Even if the true correlations with performance of overfit and underfit were the same, their correlations would differ by the ratio of their standard deviations. Given that the standard deviations may differ in empirical research, it is desirable to remove this bias before comparing the correlations of overfit and underfit. This may be done by applying the ratio of the standard deviations of overfit and underfit to correct the correlation between underfit and performance. (This is a simple correction formula, which applies unless the ratio of the standard deviations is very unequal or the correlation coefficient being corrected is large, when a more complex correction formula should be used (see Hunter, Schmidt and Jackson 1982).) The corrected correlation between underfit and performance may then validly be compared with the uncorrected correlation between overfit and performance. This provides a fair comparison that is not plagued by differential attenuation.

Once the true effects of the overfit and underfit misfits have been identified, controlling for any differences in size of misfit, or differences in standard deviation of misfit or performance, as may be required, then the analysis can proceed to test the main hypotheses. The true, relative effects of overfit and underfit can be identified. If they are the same, then the symmetry view is supported. If the effect on performance of underfit is greater (i.e., more negative) than that of overfit, then the asymmetry view is supported.

8.7.4 Managerial Design

The traditional organizational design theory, based on contingency fit theory, of symmetrical effects of misfit on performance, implicitly holds that "misfit is misfit," in that underfit is equally as bad for performance as overfit. However, the asymmetric view discussed herein holds that underfit and overfit differ in their negative effects on performance. This has implications for organizational design.

In the asymmetry view, underfit is worse than overfit, in that a unit of misfit is worse in its effect if it is under the fit line than if it is over the fit line. Underfit leads to more ineffectiveness, that is, failure to attain goals such as profit, than does overfit. Overfit leads to inefficiency, that is, unnecessary cost. However, the effect on performance of ineffectiveness is worse than the effect of inefficiency, so that underfit is worse than overfit. Therefore, an organization loses more performance from being one degree in underfit than from being one degree in overfit, over the same period of time. Hence, managers would be well advised to avoid underfit more and to act more quickly to reduce, or prevent, underfit than overfit.

When are organizations likely to be in underfit or overfit? According to the contingency theory of organizational change, that is, SARFIT (Donaldson 1987, 2001), organizations tend to be in underfit when growing and tend to be in overfit when declining (Fig. 8.8).

An organization in fit gains higher performance that leads to surplus resources, which may then be used for growth in scale, or scope (e.g., range of products/services or geographic branches). This growth increases the relevant contingency factors of the organization, such as size or diversification, while the organization retains its existing structure. Thus, the growing organization moves into misfit, in that the old structure no longer fits the new level of the contingency variable(s), for example, size or diversification. The organization is now in *underfit*, with a structure that is at an insufficient level to fit its contingency level (see Fig. 8.8). For instance, the old formalization level is now too low to fit the new size level. If size is on the horizontal axis and formalization is on the vertical axis, the fit line runs diagonally from bottom left (low size and low formalization) to top right (high size and high formalization). In this case, the organization is lying below the fit line, having too little formalization for its size, and so is in underfit. Thus, growing organizations tend to move into underfit. Hence, growing organizations are prone to underfit and so to ineffectiveness, and therefore, to more performance loss. For the period in underfit, the organization will have lower performance (other things being equal).

According to SARFIT (Donaldson 1987, 2001), an organization in fit may nevertheless decline, such as by shrinking in size because of dropping demand for its products or services. Again, an organization might have to sell one of its divisions and so become less diversified. Thus, an organization could decrease the level of its contingency factors, such as size or diversification, while retaining its existing structure. Thus, the declining organization moves into misfit, in that the old structure no longer fits the new, lower level of the contingency variable(s), for example, size or diversification (Fig. 8.8). The organization is now in *overfit*. It is lying above the fit line with a structure that is at too great a level to fit its contingency level. For instance, the old formalization level is now too great to fit the new, lower size level. Thus, declining organizations tend to move into overfit. Hence declining organizations are prone to only mild ineffectiveness, as well as to inefficiency and, therefore, to some performance loss, but not the worst – which is in underfit. For the period that they are in overfit the organization will have this low level of performance (other things being equal), but not as bad a performance as being in underfit.

Because performance loss tends to be greater from underfit than overfit, managers in growing organizations should be particularly vigilant about misfits. As contingency factors increase, managers should identify the underfits that are developing because of failure to increase structural levels. Ideally, managers and their organization designers should anticipate underfits before they occur, by increasing structural levels at the same time as increasing scale and scope, etc., so as to maintain fit. Such predictions may be made by using the body of knowledge from prior contingency theory organizational design research (Burton and Obel 1998, 2004), to give the new structural levels that would fit the degree of growth the organization is planning. Of course, ideally managers should also rapidly eliminate, or preferably avoid, overfits. But given the greater performance loss from underfits, it makes sense for managers to give priority to dealing with underfits.

A growing organization is liable to recurrently be in underfit. Each time the organization enjoys a growth spurt, it will tend to move from fit into underfit. Thus, the growing organization will experience recurrent bouts of ineffectiveness and low performance. This saps the average annual performance of the growing organization. And it thereby robs the organization of some of the surplus resources that it could have used to accelerate its growth. Thus, the recurrent periods in underfit slow the growth rate of an organization, even if it is in a highly propitious situation, such as a booming growth market. Therefore, the episodes of underfit are a drag on the growing organization that prevents its full acceleration. Hence, managers in a growing organization would be well served to minimize underfits, by quickly adjusting out of them or preferably avoiding them. This gives a premium to the organizational design knowledge (Burton and Obel 1998, 2004) that helps managers and consultants escape, or avoid, underfits, as their organization grows. Otherwise, episodes of underfit have the potentiality to be the Achilles heel of organizational growth. They are vulnerable weak spots, which, even if not fatal, hobble the organization in its race for growth. They could be fatal, if the ineffectiveness from underfit happened to be accompanied by a downturn in other causes of organizational performance, such as a slump in sales. Overall, underfit, by under-cutting growth, means that bright prospects are tarnished.

8.8 Conclusions

Traditional organizational design based on structural contingency theory holds that underfit reduces performance the same as overfit, so that the effects of misfit are symmetrical. In contrast, the asymmetric theory holds that, while both underfit and overfit reduce performance, their effects differ in magnitude. Specifically, Klaas et al. (2006) hold that underfit produces ineffectiveness, that is, failure to attain goals, while overfit produces inefficiency, that is, unnecessary cost of structure. Thus, underfit produces worse performance than overfit, that is, asymmetry. This posits that cost from structure rises with the level of the structural variable, while benefit from structure rises until fit to the contingency is reached, when benefit plateaus.

Formal modeling herein identifies the relationships more exactly, showing how underfit is worse for performance than overfit. Thus, as the structural level initially increases, cost and benefit rise, with benefit rising more steeply, producing increased performance as underfit reduces. After fit is reached, further increases in structure produce increasing overfit, which has increased cost, while benefit remains constant, so that performance decreases as overfit increases. The extent to which benefit exceeds cost is greater in overfit than underfit, so that performance is lower for underfit than overfit. Thus, underfit causes increasing ineffectiveness, while overfit causes increasing inefficiency.

Asymmetric effects of misfits on performance underlain by the rising cost of structure imply that the performances of fits to higher levels of the contingency are lower than the fits to lower levels of the contingency. However, for many contingency variables, it is likely that fits to higher levels of the contingency produce higher performance than fits to lower levels. This condition can be met if the contingency variable (e.g., size) contributes positively to performance. This positive effect of the contingency on performance is a main effect that accompanies the contingency effect, but is separate from it. Such a positive main effect of contingency on performance is postulated for intra-organizational contingencies. However, for extra-organizational contingencies, asymmetry could produce declining performances of fits, with no off-setting positive main effect of contingency on performance.

There is a need to empirically investigate asymmetry in future research to validate the ideas herein by testing between the competing theories of symmetry and asymmetry. A methodological consideration is that comparisons of underfits and overfits must be careful to avoid possible confounds. In particular, comparisons of the mean performances of underfits and overfits must control for any differences in the degree of misfit between the two types of misfit. Similarly, comparisons of the effects of underfits and overfits, using regression, or correlation, coefficients, must control for any differences between the two types of misfit that are due to differences in standard deviations of either misfit or performance. This can be accomplished by correcting for differences due to the standard deviations of underfits and overfits before comparing their regression or correlation coefficients.

Managers and organization designers would be well advised to seek to eliminate and prevent both underfits and overfits. However, given the more severe impact on performance of underfit than overfit, managers should give priority to eliminating underfits. Underfit is more likely to occur in growing organizations, so managers and organization designers should be on the look out for underfit in that situation, where underfit causes a drag on the rate of growth. However, it may also be the case that overfit is more common than underfit, and that misfit is on average more severe for overfit than underfit, so that the effect of overfit is still important economy-wide.

References

Bertalanffy L v. (1968) General systems theory: Foundations, development, applications, 2nd edn. George Braziler: New York.

Burton R, Lauridsen J, Obel B (2002) Return on assets loss from situational and contingency misfits. Management science 48 (11): 1461–1485.

Burton R, Lauridsen J, Obel B (2004) The impact of organizational climate and strategic fit on firm performance. Human Resource Management 43 (1): 67–82.

Burton R, Obel B (1998) Strategic organizational diagnosis and design: Developing theory for application, 2nd edn. Kluwer: Boston.

Burton R, Obel B (2004) Strategic organizational diagnosis and design: The dynamics of fit, 3rd edn. Kluwer: Boston.

Child J (1972) Organizational structure, environment and performance: The role of strategic choice. Sociology 6: 1–22.

Child J (1975) Managerial and organizational factors associated with company performance, Part 2: A contingency analysis. Journal of Management Studies 12: 12–27.

Donaldson L (1987) Strategy and structural adjustment to regain fit and performance: In defense of contingency theory. Journal of Management Studies 24 (1): 1–24.

Donaldson L (1996) Structural contingency theory. In: Clegg S, Hardy C, Nord W (eds), Handbook of organization studies. Sage: London, pp 57–76.

Donaldson L (2001) The contingency theory of organizations. Sage: Thousand Oaks, CA.

Doty D, Glick W, Huber G (1993) Fit, equifinality and organizational effectiveness. Academy of Management Journal 38 (8):1196–1250.

Drazin R, Van de Ven A (1985) Alternative forms of fit in contingency theory. Administrative Science Quarterly 30: 514–539.

Galbraith J (1974) Organization design: An IP view. Interfaces 4 (3): 28–37.

Galbraith J (1977) Organization design. Addison-Wesley: Reading, MA.

Grinyer PH, Yasai-Ardekani M (1981) Strategy, structure, size and bureaucracy. Academy of Management Journal 24 (3): 471–486.

Hunter JE, Schmidt FL, Jackson GB (1982) Meta-analysis: Cumulating research findings across studies. Sage: Beverly Hills, CA.

Keller RT (1994) Technology-information processing fit and the performance of R&D project groups: A test of contingency theory. Academy of Management Journal 37: 167–79.

Klaas P, Lauridsen J, Håkonsson DD (2006) New developments in contingency fit theory. In: Burton RM, Eriksen B, Håkonsson DD, Snow CC (eds), Organizational design: The evolving state-of-the-art. Springer Science and Business Media: New York.

Meyer A, Tsui A, Hinings C (1993) Configurational approaches to organizational analysis. Academy of Management Journal 38 (6):1175–1195.

Miller D, Friesen P (1980) Momentum and revolution in organizational adaptation. Academy of Management Journal 23: 591–614.

Naman J, Slevin D (1993) Entrepreneurship and the concept of fit: A model and empirical results. Strategic Management Journal 14:137–153.

Scott R (2003) Organizations, 5th edn. Prentice Hall: Upper Saddle River, NJ.

Simon H (1982) The sciences of the artificial, 2nd edn. MIT Press: Cambridge, MA.

Tushman M, Nadler D (1978) IP as an integrating concept in organizational design. Academy of Management Review 3: 613–624, July.

Van de Ven AH, Drazin R (1985) The concept of fit in contingency theory. In: Staw BM, Cummings LL (eds), Research in Organizational Behaviour, Vol. 7. JAI Press: Greenwich, CT, pp 333–365.

Venkatraman N (1989) The concept of fit in strategy research: Toward verbal and statistical correspondence. Academy of Management Review 14 (3): 423–444.

Woodward J (1965) Industrial organization: Theory and practice. Oxford University Press: London.

Zajac E, Kratz M, Bresser R (2000) Modelling the dynamics of strategic fit: A normative approach to strategic change. Strategic Management Journal 21: 429–453.

Index

A
Adaptation, 43–59, 68, 81, 82, 87, 124, 125, 126, 127–128, 131, 132, 133, 137, 140, 142, 160

C
Capability/capabilities, 4, 8, 14, 15, 16, 43, 45, 49, 51, 55, 61–76, 80, 82, 84, 91, 93, 111, 141
Change, 37, 41–117, 123, 124, 125, 126, 127, 128–130, 132, 133, 134, 137, 142, 157, 161, 164
Collaboration, 5, 13, 14, 15, 17, 29, 127
Collaborative community, 3–18
Collaborative innovation networks, 4–9, 10, 13–14
Community of firms, 3–18
Configuration, 13, 79–93, 148
Contingency theory, 35, 36, 80, 81, 82, 84, 85, 91, 147–167
Coordination, 15, 25, 31, 45, 54, 55, 56, 57, 58, 59, 82, 85, 87, 88, 100, 101, 102–107, 108, 109, 110, 114, 115, 116, 126, 132, 136, 137

D
Design, 6, 10, 16–17, 29–34, 43, 44, 45, 55–56, 57, 61–76, 91–93, 99–117, 126, 130, 131, 132, 133, 141, 147–167
Design rules, 62, 68, 69, 70–75
Dynamic capability/dynamic capabilities, 43, 44, 45, 51, 57, 59, 61–76, 80, 82, 84, 91, 92

F
Fit, 121–167

I
Implementation, 18, 45, 123–143
Information processing theory, 107, 115, 116, 155–156
Innovation, 3, 4–9, 10, 13–14, 15, 16, 17, 28, 32, 33, 34, 50, 70, 71, 81, 84, 87, 88, 91, 123–143
Institutional theory, 124
ISO 9000, 130–131, 132, 133, 134, 135, 140, 141, 142

L
Learning, 5, 27, 28, 33, 66, 71, 72–73, 74, 108, 123–143

M
Misfit, 82, 114, 116, 147–167

N
Network(s), 5, 16, 23–37, 64, 93, 102, 105, 128
design, 23–37
governance, 24–26, 29, 34, 35, 37
tasks, 26, 27, 28, 30, 33, 34
theory, 23–37
New organizational forms, 1–37

O
Open source, 5
innovation, 4, 5, 9
Organizational change, 70, 71, 73, 75, 79, 89, 125, 127, 164
Organizational design, 10, 36, 45, 46, 50, 53, 54, 55–56, 57, 59, 68, 75, 99, 100, 103, 116–117, 147–167
Organizational dynamics, 63, 65, 66, 68
Organizational environments, 43, 44, 56, 57, 61, 63, 68, 73, 74, 80, 81–83, 91, 103, 109, 148
Organizational forms, 1–37, 53, 82, 87, 88, 89
Organizational tradeoffs, 43–59
Organization design, 16–17, 37, 43, 44, 46, 48, 51, 56, 57, 59, 62, 85, 91–93, 99–117, 141, 147, 148, 165, 167

Organization design (*cont.*)
research, 101, 103, 116
theories, 4, 16–17, 44, 91, 99–117, 141
Organization theories, 79, 80, 141, 161–166
Overfit, 147–167

P
Performance, 29, 48, 53, 54, 55, 56, 58, 62, 63, 65, 66, 67, 69, 70, 71, 74, 76, 81, 82, 83, 84, 85–86, 89, 90, 91, 92, 105, 123–143, 147–167
Population ecology theory, 91
Processes, 4, 5, 7, 8, 9, 10, 11, 13–14, 15, 16, 17, 18, 26, 27, 28, 29, 32, 33, 46, 47, 48, 51, 52, 53, 54, 58, 76, 80, 91, 93, 108, 110, 111, 112, 114, 124, 125, 127, 128, 130, 131, 135, 140, 141, 142, 156, 158

R
Reconfiguration, 79–93
Redesign, 45, 52
Reorganization, 80, 123
Resource dependence theory, 91
Restructuring, 80, 81, 82, 85, 86, 89
Routines, 16, 44, 62, 63, 64, 65, 66, 67, 69, 70, 71, 72, 73, 74, 75, 76, 87, 89, 92, 104, 105, 124, 125, 128, 129, 130, 131, 133, 141, 142

S
Search, 26, 44, 45, 49, 53, 56, 57, 58, 59, 71, 73, 81, 86, 90, 112, 127, 128, 129, 141, 142
Strategic choice, 83, 94
choice theory, 92
Structural change, 44, 79, 80, 82, 84, 85, 86, 88, 91, 92, 93
Structure, 9, 11–13, 17, 25, 31, 33, 34, 35, 36, 37, 44, 45, 46, 47, 49–55, 56, 57, 64, 73, 79, 81, 83, 85, 87, 88, 90, 91, 92, 93, 105, 129, 147, 148, 152, 153, 155, 156, 160, 164, 165

T
Threat-rigidity theory, 91
Transaction cost theory, 85

U
Underfit, 147–167

V
Virtuality, 99–117
Virtual organizations, 99–117